DEVELOPMENTS IN SEDIMENTOLOGY 42

CARBONATE – CLASTIC TRANSITIONS

DEVELOPMENTS IN SEDIMENTOLOGY 42

T.M.

CARBONATE – CLASTIC TRANSITIONS

Editors

LARRY J. DOYLE

Department of Marine Science, University of South Florida, 140 7th Ave. South, St. Petersburg, FL 33701 (U.S.A.)

and

HARRY H. ROBERTS

Louisiana State University, Coastal Studies Institute, Baton Rouge, LA 70803 (U.S.A.)

ELSEVIER
Amsterdam — Oxford — New York — Tokyo   1988

ELSEVIER SCIENCE PUBLISHERS B.V.
Sara Burgerhartstraat 25
P.O. Box 211, 1000 AE Amsterdam, The Netherlands

Distributors for the United States and Canada:

ELSEVIER SCIENCE PUBLISHING COMPANY INC.
52, Vanderbilt Avenue
New York, NY 10017, U.S.A.

ISBN 0-444-42904-2 (Vol. 42)
ISBN 0-444-41238-7 (Series)

Printed in The Netherlands

V

PREFACE

Over the years the field of Sedimentology has become subdivided into various specialities. Two of the largest groups are made up of those who study clastic rocks and those who study carbonates. There seems to be less and less communication between them. Whole journals have sprung up exclusively for one or the other and research conferences and groups tend to be mutually exclusive. Of course, the rocks are ignorant of our pigeon holes and the facies change from clastic to carbonate both laterally and through time. This volume stems from our observations of these kinds of changes in the Gulf of Mexico and our realization that these geologically important transitions were being largely ignored because of our professional compartmentalization.

The book that follows is made up of 11 chapters. We begin with a chapter by Hay _et al_. which provides an overview of the big picture of global patterns of carbonate and clastic sedimentation. We then proceed to a discussion of sedimentary models of siliciclastic deposits and coral reef relationships by Santisteban and Taberner. There follow eight case studies of carbonate clastic transitions and a concluding chapter by Murray _et al_. on control of carbonate-clastic sedimentation systems by baroclinic coastal currents.

In addition to presenting what we feel are a group of excellent research papers, our goal with this volume is to emphasize that clastic and carbonate sedimentation are not separate but part of a continuum; a transition that needs to be more thoroughly investigated and better understood.

We would like to express special thanks to Mark Sherwin who provided valuable contributions in technical editing and preparation to this volume.

Larry J. Doyle

Harry H. Roberts

Table of Contents

Chapter 1

PLATE TECTONIC CONTROL OF GLOBAL PATTERNS OF DETRITAL AND CARBONATE SEDIMENTATION

W.W. HAY, M.J. ROSOL and J.L. SLOAN II
University of Museum, Boulder, CO   80309
D. E. JORY
Rosenstiel School of Marine and Atmospheric Sciences, University of Miami, Miami, FL   33149

ABSTRACT

Global patterns of continental drainage to the oceans have changed markedly over the last 200 m.y. in response to plate tectonic processes; most of the earth's major rivers now enter the sea on passive continental margins which did not exist in the early Mesozoic. This reorganization of drainage has strongly influenced the distributions of marine detrital and carbonate facies.

Analysis of changes in continental topography related to the breakup of Pangaea suggest that throughout much of the Mesozoic, drainage systems were dominated by a pole-to-pole divide directing detrital sediment away from the sites of future continental rifting. This phase was followed by rifting and formation of narrow oceans with uplifted margins. As the margins subsided by thermal relaxation, massive amounts of detrital sediment were delivered from the continental interiors onto the young passive margins. In time, river drainage became increasingly focused, concentrating detrital sediment supply at the mouths of a few large rivers. Very large supplies of detrital sediment require large, high uplifts such as those caused by subduction of young, hot ocean crust or by continental collision.

Large sediment supplies also require drainage basins with relatively constant slope; so that sediment erosion, throughput, and delivery to the ocean margin are efficient. The result is rapid sedimentation of deltaic complexes containing an abundance of organic carbon. During most of earth history, there are no large, high uplifts, and carbonate rocks become more important in the continental margins.

In contrast to the point inputs of detrital sediments, the supply of carbonate has been from the oceanic reservoir and is diffuse. Carbonate deposition dominates the continental shelves in all warm regions where the detrital sediment input is not extremely large. Carbonate shelves become cemented, resisting erosion, so they build up until the shelf edge approximates highstands of sea level. Detrital shelves become adjusted to lowstands of sea level with the shelf breaks typically many tens of meters below the low sea level.

The clastic-carbonate shelf-slope-rise system operates to promote bypassing of detrital materials into deep water in the subtropics and tropics, with sharp facies contrasts. In higher latitudes, carbonate may be a significant proportion of the continental margin material, but facies changes are usually much more gradual.

1 INTRODUCTION

The key to developing and understanding global patterns of sedimentation is a global inventory of sediment types. This sort of inventory had been attempted by Poldervaart (1955) and Ronov (1959, 1964) but because of our lack of knowledge of the sediments presently beneath the sea, their estimates were largely based on extrapolations of information derived from studies of rocks on land.

Garrels and Mackenzie's (1971) classic book "The Evolution of Sedimentary Rocks," provided both new estimates of the distribution of sedimentary rocks in space and time and a framework in which to predict composition and quantity of marine deposits. Hay and Southam (1977) analyzed the arguments presented by Garrels and Mackenzie in terms of oceanic sedimentation modulated by the dynamics of plate tectonics, with sedimentation on the continental shelves responding to thermal subsidence and sea level changes. They concluded that on geologically short term (10^5 years) time scales, perturbations of supply of material to the ocean could be effected by changing sea levels and shelf deposition or erosion.

However, it was evident that on longer time scales, the supply of sediment to the sea may far exceed the capacity of the continent shelves as a sediment site. The recent supply of sediment by rivers so vastly exceeds the space available for sedimentation on the shelves that almost 90% of it must be bypassed into the ocean proper. On a continental scale, the dissolved load of rivers appeared to be a simple linear function of the area of the continent draining to the sea. The detrital load of rivers appeared to be linear function of area available for erosion and a non-linear, apparently exponential, function of the elevation of the continent. Using these assumptions, Southam and Hay (1977) made a prediction for the changes in chemical input to the ocean for the past 200 m.y.; Southam and Hay (1976) had also explored Broecker's kinetic model for the composition of seawater using dynamical formulation. They concluded from the shape of the curves for oceanic response to changes in river input and to vertical mixing processes, that it was unlikely that changes in vertical mixing process affecting fertility would be detectable in sedimentation rates averaged over several million years, but might be very important on shorter time scales. Because both sea level changes affecting the continental shelf and internal mixing processes modulate the supply of materials in the ocean only on time scales in the order of hundreds of thousands of years and because Pliocene and older strata are routinely dated by biostratigraphic and paleomagnetic schemes having resolution of about two million years, Southam and Hay (1977) concluded that the measurable fluctuations in the accumulation rate of $CaCO_3$ in the ocean should not vary by more than 50% in the past 100 m.y.

The Deep Sea Drilling Project (DSDP) has been unique among studies in the development of geology because it has produced the data necessary to consider the earth a closed system. Until a large quantity of information became available concerning sedimentation in the deep sea, geological studies focussing on smaller areas could always draw on supplies of material from outside the area and could lose material to the rest of the world. The DSDP has made it possible to consider the world in terms of mass balances. The DSDP data set is the largest coherent body of information on sedimentary mass balance; it is inherently important for all studies of global distributions and mass balance calculations. The following is a summary account of how knowledge of the integrated data base was developed and the major trends it indicates.

From cursory examination of the DSDP data from early legs of the project, it became evident that sedimentation rates in individual ocean basins had varied markedly with time. It was assumed in unpublished studies that these variations were largely a result of partitioning between the major ocean basins. Davies, Worsley, Hay and Southam set about determining average sedimentation rates in each of the ocean basins to explore partitioning. The result was the discovery (Davies et al., 1977) that sedimentation rates had apparently varied synchronously in all ocean basins by a factor of five, with high rates during the Mid Eocene and Mid Miocene-recent and low rates during the Paleocene-Early Eocene and Late Eocene-Early Miocene. These changes in oceanic sedimentation rates appear to be global and to reflect real changes in the supply of materials from land. Davies et al. suggested that during times of low overall sedimentation rates continental climates might have been arid, similar to Australia today. It was recognized that for younger sediments, sedimentation rates, which include pore space, may be significantly different from accumulation rates, which measure only the solid phase. The extremely high values for Pleistocene-Pliocene sedimentation rates were assumed to be in part a function of the large amounts of pore space in these deposits in the deep sea. Investigation of unconformities and hiatuses has demonstrated regional correlations which indicated changing basinwide sedimentation patterns (Rona, 1973, for the North Atlantic; Davies et al., 1975, for the Indian Ocean).

Moore and Heath (1977) documented the global abundance of hiatuses for the Cenozoic in the world ocean. They found that the linear recycling model of Garrels and Mackenzie (1971) overestimates the proportion of surviving Neogene sections and underestimates those of the Paleogene. In particular, unusually high proportions of section represented by hiatus were noted for the early

Paleocene and the Late Eocene-Early Oligocene; these are the sedimentation rate
minima noted by Davies et al. (1977). They found that these deviations could be
removed by assuming the mass of sediment deposited per time increment might
vary by a factor of four; this is essentially variation of the supply.
Alternatively the data could be fit by varying the chance of erosion by a
factor of twenty. Using both effects together, they obtained a good fit by
varying each of these parameters by lesser amounts.

Berger (1979) has noted the similarity of the Moore and Heath (1977) and
Davies et al. (1977) curves. The close correlation is not unexpected because
they are two different forms of expression of what is or is not present in each
ocean basin. They were, however, constructed from different kinds of data and
the two sets of information analyzed were not identical but only overlap in
part.

More detailed analysis of Cenozoic hiatuses has been carried out by Moore et
al. (1978) who have interpreted the patterns in terms of the deep circulation
history of the ocean.

Simple averaging of sediment thickness for accumulation rate data or of
abundance of hiatuses is a valid technique for determining rates for basins or
for the world ocean only if the data are a representative sample of the basins
or global ocean. Fortunately, the objectives of core recovery by the DSDP were
so diverse that the resulting collection of cores is a near random sample. The
distribution of cores for early Legs was discussed by Moore (1972) and for the
first 335 sites by Whitman and Davies (1979). The Northern Hemisphere is
overrepresented, as are sites in water depths between 1 and 3 km. The Southern
Hemisphere and sites in water depths less than 1 km are underrepresented.
Continental rises, arcs and trenches and sea mounds and aseismic ridges are all
overrepresented; shelves, ocean basins and the mid-ocean ridge system are
somewhat underrepresented.

Worsley and Davies (1979a) have presented a more detailed account of deep
sea sedimentation using accumulation rates averaged over intervals of equal
time length (3 m.y.). Davies et al. (1977) had determined sedimentation rates
by averaging sediment thicknesses over classical chronostratigraphic intervals
(Late Miocene, Middle Eocene, Early Paleocene, etc.) of unequal time length.
The effects of using accumulation rates is to damp the younger part of the
curve; the effect of using time intervals of equal length is to generally damp
the magnitude of fluctuations.

Worsley and Davies (1979a) compared the new global sediment accumulation
rate patterns with the sea level curve of Vail et al. (1978) and found a good
correlation, confirming the idea that eustasy plays an important role in

determining whether sediment accumulates on the continental shelves or in the deep seas.

Worsley and Davies (1979b) have recently embarked on a more sophisticated method of exploring the patterns and rates of sediment accumulation in the ocean, plotting a series of rate and lithology maps for the Pacific. Their analysis of the Pacific extends back only to the Middle Eocene, but shows a Middle Eocene high rate, a late Eocene-Middle Oligocene low rate, late Oligocene-early Miocene high rate, a middle and late Miocene low rate, and very high rates for the Pliocene and Pleistocene. Again, the total variation in accumulation rates is through a range of a factor of four. Davies and Worsley (1981) calculated the global land-sea flux over the past 60 m.y. to average $10.3-12.5 \times 10^{14}$ g/yr, values very close to the 12.2×10^{14} g/yr estimated for the present day. They found that oceanic accumulation rates varied between 7.8 and 28.6×10^{14} g/yr and concluded that the high rates of carbonate accumulation in the deep sea correspond to times of maximum exposed continental area, i.e. low stands of sea level and low accumulation rates in the deep sea to high stands of sea level. Using the algorithm of Davies et al. (1977), Southam and Hay (1982) have produced global curves for total sedimentation rate, carbonate sedimentation rate, non-carbonate sedimentation rate and organic carbon sedimentation rate for the Atlantic, Pacific and world oceans, back to the mid Cretaceous, confirming the fluctuations in sedimentation rates previously noted, and determining that as other components vary through factors four or five, organic carbon sedimentation rates vary through several orders of magnitude.

All of these studies of sedimentation in the world's ocean basins have resulted in our knowledge of the patterns of distribution of oceanic sediments through time surpassing our knowledge of patterns of sediment distribution on the shelf seas and on continental margins through time. Although a few offshore areas have been extensively drilled, knowledge of continental margin sediments is generally poor, especially with regard to rates of sedimentation.

The purpose of this paper is to explore and predict distributions of carbonates and clastics in continental margin settings, based on a new understanding of the basic new principals governing sediment delivery to the sea and partitioning of sediment types between the shelf, slope and rise, and deep sea. Our relatively detailed knowledge of what has happened in the deep sea provides a background for understanding what must have been happening along the continental margins.

2 THE PRESENT DISTRIBUTION OF SEDIMENTARY MATERIALS

Southam and Hay (1982) have estimated the areas, volumes and masses of major sedimentary reservoirs (figure 1). Paleozoic and older sediments are essentially restricted to the craton and PreCambrian-Paleozoic geosynclines. The remainder of the sediments are Mesozoic-Cenozoic. The shelves, slopes and rises of the passive margins which developed as a result of the breakup of Pangaea are the largest single body of Mesozoic-Cenozoic sediment. The next largest reservoir is the Mesozoic-Cenozoic geosynclinal deposits. The pelagic sediments form a distinctive unit, large in volume and quite different from the other reservoirs in chemical composition in being unusually enriched in $CaCO_3$ since 100 Ma (Southam and Hay, 1977). The total volume of sediment on the continental blocks is $674x10^6 km^3$ according to Southam and Hay (1982), which is not greatly different from Ronov's (1982) estimate of $758x10^6 km^3$. The effect of the breakup of Pangaea has been to recycle two-thirds of the total sedimentary mass, with more than one-half of this recycled mass being offloaded from the continental blocks onto oceanic crust generated during the Mesozoic and Cenozoic.

3 THE SUPPLY OF SEDIMENT TO THE OCEANS

The major suppliers of sediment to the oceans are the rivers. The river flux of dissolved material is approximately $4.1x10^{15}$g/yr including $0.2x10^{15}$g/yr of organic matter. The detrital load of rivers is in two components; the suspended load, estimated by Garrels and Mackenzie (1971) to be about $18.2x10^{15}$g/yr, and the bed load estimated by Holland (1978, 1982) to be about 10% of the suspended load, or about $1.8x10^{15}$g/yr. The total detrital load of rivers is thus $20.0x10^{15}$g/yr and total detrital and dissolved load carried by rivers is about $24.1x10^{15}$g/yr. Glaciers are the second most important source of sediment to the sea, but carry only detritus, about $2.0x10^{15}$g/yr (Garrels and Mackenzie, 1971). The dissolved load carried by the groundwater ($.5x10^{15}$g/yr according to Garrels and Mackenzie, 1971) is about one tenth the dissolved load of the rivers, and the eolian flux, recently reestimated by Prospero (1982) is .5 to $.8x10^{15}$g/yr; both are unimportant in terms of global balances. Supply by glaciers has probably only been significant during the late Cenozoic, so that for the interval between the end of the Paleozoic and the middle Miocene rivers were the only important suppliers of sediment. In exploring the global distributions of carbonate and terrigenous detrital sediment in the context of plate tectonics, it is important to know what controls the relative proportions of dissolved and detrital loads of rivers.

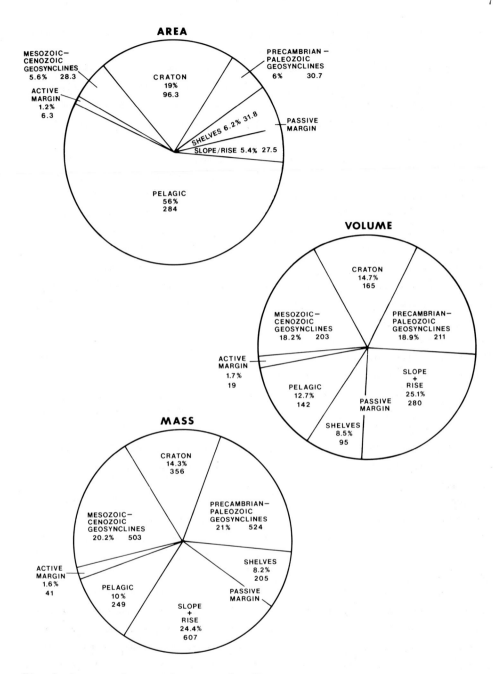

Fig. 1. Areas, volumes and masses of sediment present on earth today in major tectonic reservoirs. Only the craton and Precambrian-Paleozoic geosynclines are pre-Mesozoic. Numerical values are in $10^6 km^2$ for area, $10^6 km^3$ for volume and 10^{21}g for mass. From data of Southam and Hay, 1981.

The type of rock exposed in the drainage basins may significantly affect the dissolved load of rivers. The mass of dissolved material is lowest for rivers draining plutonic igneous rock terranes; rivers draining volcanic areas carry about twice as much, and those draining sedimentary terranes about ten times as much dissolved material (Lerman, 1982). Rivers with high sulfate and chloride concentrations invariably drain basins which include near surface or exposed evaporite deposits (Holland, 1978, 1982) and are recycling salts.

The concentration of dissolved silica is a function of weathering processes relative to the mean temperature of the drainage basin, being significantly higher in tropical rivers (Hay and Southam, 1977; Lerman, 1982).

Assuming that all of the dissolved Ca in rivers becomes $CaCO_3$ in the sea, the present flux of $CaCO_3$ to the sea is 1.24×10^{15} g/yr (Hay and Southam, 1977). Assuming that all of the CO_2 in the sedimentary reservoirs (Table 2 of Hay in press) is in the form of $CaCO_3$, the mass of $CaCO_3$ in Mesozoic-Cenozoic sediments is 425.7×10^{21} g. It follows that the average rate of supply of $CaCO_3$ to the ocean must have been 1.72×10^{15} g/yr over the past 248 m.y. or about 40% higher than the present rate of delivery. This is in spite of the fact that the land area exposed to weathering and draining to the sea is now at a maximum. This results from the fact that at the end of the Paleozoic the continents had a much thicker sediment cover, a large portion of which was carbonate, and that the calcareous plankton appeared in abundance 100 m.y. ago, and have been removing $CaCO_3$ from the continent to shelf sea cycle into deep sea sediments ever since (Hay and Southam, 1976; Sibley and Vogel, 1976; Southam and Hay, 1977; Sibley and Wilband, 1977). Because of the loss of the supply of $CaCO_3$ at its source and its removal into the deep sea, less and less $CaCO_3$ is available on the continents to enter the river systems.

The detrital load of rivers is less well understood because of the greater difficulty in measuring it, because of the special significance of major floods, and because of the effects of man and his agricultural practices (Garrels and Mackenzie, 1971; Hay and Southam, 1976; Holland, 1978; Milliman, 1974).

Hay (1983) presented the National Center for Atmospheric Research (NCAR) 5° square average elevation map with a contour interval of 1 km and indicating the drainage basins of 29 important rivers. The map shows only six areas over 1 km elevation, two of these are ice caps, Antarctica and Greenland, and the other four are western North America, western South America, eastern Africa, and the Himalayan-Tibet region; each of these four uplifts reflect major plate tectonic processes, as will be discussed below. It is very significant that no rivers cross these four uplifts, but 24 of the 29 most important rivers have

their headwaters on the uplifts, and the drainage systems radiate from the
uplifts.

The river carrying the largest detrital load is the Huang Ho (Yellow) which
delivers about 1.9×10^{15} g/yr or 10% of the global detrital load, to the Yellow
Sea. However, the Huang Ho is a special case as most of its detrital load is
derived from erosion of the loess area of China which has probably been
strongly affected by agricultural practices over the past few millennia. The
present load of the Huang Ho is an abnormal fluctuation in the long term
process, delivering material previously stored in the drainage basin.

Discounting the Huang Ho, the major rivers draining the Himalayan-Tibet
uplift to the south and east, the Indus, Ganges, Brahmaputra, Irrawady, Mekong,
and Yangtze, deliver a combined total of 3.6×10^{15} g/yr (compiled from Table 7 of
Lisitzin, 1974) or 20% of the world's total detrital load. This enormous load
is derived from drainage basins totalling only 5.8×10^{6} km^{2}. With the Huang Ho,
the load is 30% of the global total derived from an area of 6.5×10^{6} km^{2}. Hence,
one-fifth to one-third of the global detrital load is derived from only 3% of
the land area.

Lisitzin (1974) indicated that the 55 rivers having the largest detrital
loads supply 8.1×10^{15} g/yr from drainage basins having an area of 55×10^{6} km^{2}, or
45% of the total detrital supply to the oceans is derived from 36% of the land
area. More significantly 30 of these rivers have headwaters on one of the four
major uplifts; they carry 7.6×10^{15} g/yr from drainage basins having an area of
43×10^{6} km^{2}, or 42% of the total detritus from 28% of the land area. It is
evident that large uplifts play a major role in determining supply of detrital
sediment to the sea, but climate and geology of the drainage basin may also be
significant factors.

Figure 2 shows the relationship between average elevation of a number of
major drainage basins and the detrital erosion rate of each basin, as
determined by dividing the annual suspended load by the area of the basin.
Regression lines fitted to the data have standard deviations too large to
permit the equations to be used as anything more than a guide to approximate
the detrital load. It is evident that details of the hypsography of the
drainage basin are essential to understanding the delivery of detritus to the
sea. Figures 3-6 show the hypsography of several major drainage basins, with an
indication of the sediment load carried by each river. It is evident that there
is generally a close relationship between the hypsography of the drainage basin
and the detrital load of the river, but several peculiarities should be noted.

The very high detrital loads are associated with basins having steep
hypsographic curves, and there is even a strong suggestion that for basins such

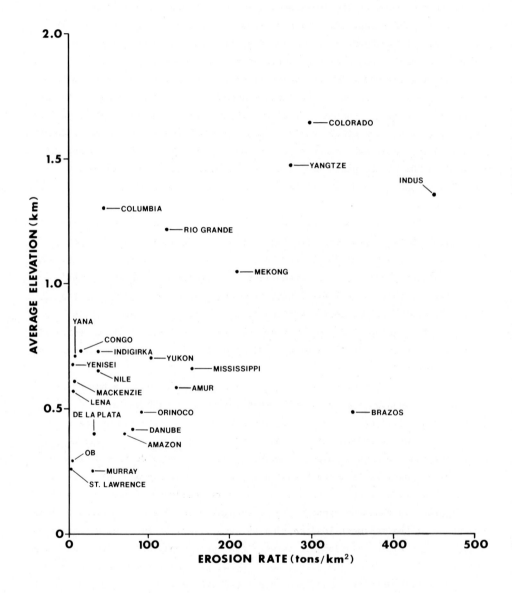

Fig. 2. Detrital erosion rate (after Lisitzin, 1974) plotted against average elevation of twenty-four major drainage basins.

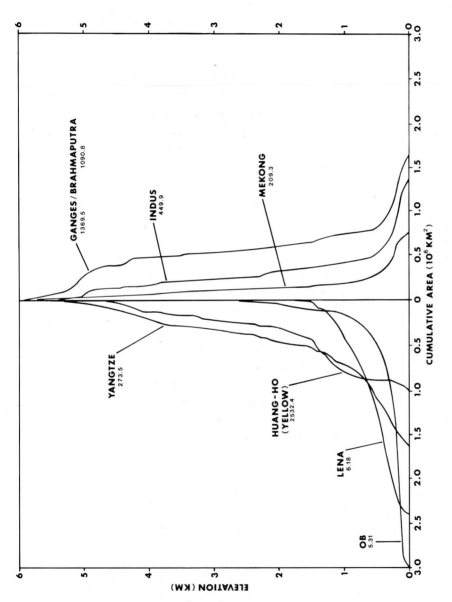

Fig. 3. Hypsographic curves for seven major river basins draining the Himalayan/Tibetan uplift. Detrital erosion rate of the basin is given in metric tons/km².

as the Ganges-Brahmaputra and Indus (fig. 3) there may be significant
aggradation in the lower part of the basin. Such very steep hypsographic curves
for major rivers are generally restricted to the Himalayan-Tibetan uplift, a
product of continental collision, but the upper reaches of the Amazon basin are
similar (fig. 4). The Mississippi basin presents a more typical broadly concave
hypsographic curve (fig. 5). Some of the other hypsographic curves are more
irregular, with flat areas in the upper parts of the drainage basin suggesting
local sites of aggradation and sediment storage. Rivers with steep slopes in
the lower part of the hypsographic curve, such as the Zambesi (fig. 6) supply
anomalously high detrital loads to the sea, suggesting that sediment throughput
is complete, and little if any significant aggradation occurs. Large drainage
basins with hypsographic curves having relatively constant slope will result in
efficient sediment throughput and have a good chance of producing deltaic
complexes. Because sedimentation rates are high, high organic carbon contents
can be expected (Southam and Hay, 1981). The detrital load of rivers is clearly
a function of elevation and hypsography of the drainage basin, and is
undoubtedly related to climatology and geology as well. Because the space
available for sedimentation of carbonate in epeiric seas and on the continental
shelves is a function of the space not occupied by detrital sediments, the key
to shallow water carbonate deposition lies in understanding and predicting how
the detrital sediment supply may have varied in the past.

4 RECONSTRUCTING PALEOTOPOGRAPHY OF THE CONTINENTS

It is generally accepted that the bathymetry of the ocean basins reflects
the distribution of heat in the upper mantle and ocean crust. As the
lithosphere cools, it becomes more dense, so that sea floor bathymetry can be
approximated as a function of the age of ocean crust. The controls on
topography of the continental areas may be only slightly more complex,
depending on the thickness of the continental crust as well as distribution of
heat in the upper mantle. Figure 7 is an adaptation of the NCAR average
elevation map with elevations averaged over 5° squares, with a 1 km contour
interval (Hay, 1983). It also shows dominant shelf sediment type for the 5°
squares. There are at present only six areas with elevations above 1 km on this
map; two of these areas are ice caps: Greenland and Antarctica; one is the
area of very thick (50-80 km) continental crust underlying the Himalayas and
Tibet (Soller et al., 1982), resulting from collision of two continental
blocks; two areas are related to subduction of young, hot, light oceanic
lithosphere: western South America and western North America; and one area,
extending from northeastern to southwestern Africa, is most if not entirely

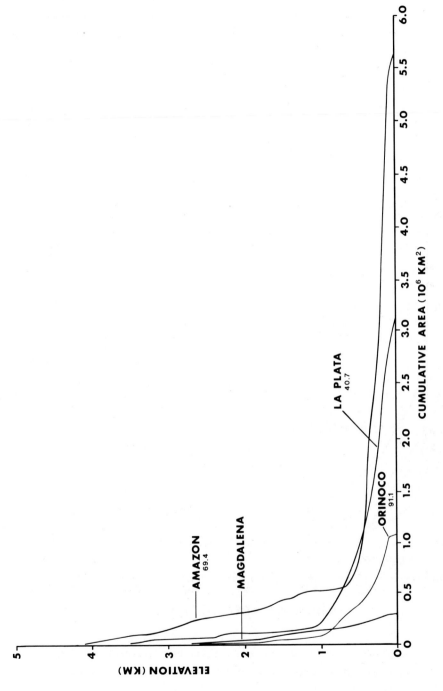

Fig. 4. Hypsographic curves for four major river basins draining the Andean uplift. Detrital erosion rate of the basin is given in metric tons/km², where known.

14

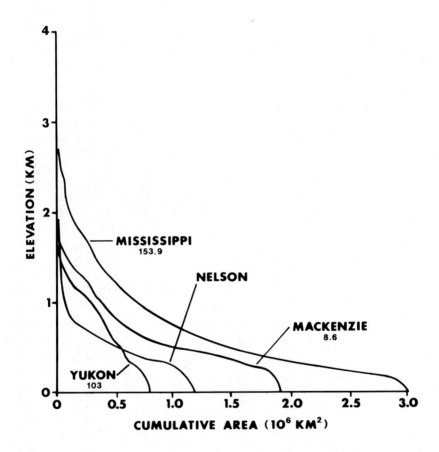

Fig. 5. Hypsographic curves for four major river basins draining western North America. Detrital erosion rate of the basin is given in metric tons/km^2, where known.

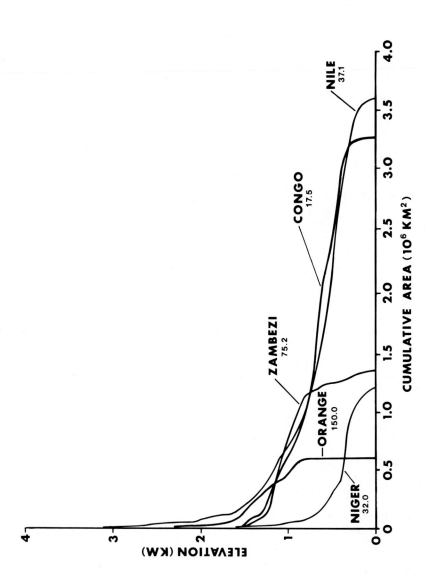

Fig. 6. Hypsographic curves for five African river basins. Detrital erosion rate of the basin is given in metric tons/km².

16

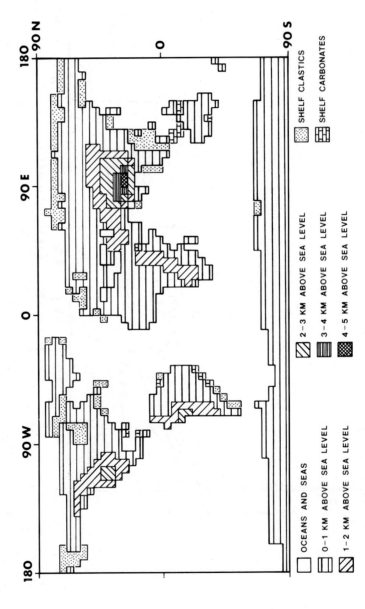

Fig. 7. Average elevation map for 5° squares with 1 km contour interval. Shelf and epeiric seas and their dominant sediment type for 5° squares are also shown. Major predictable drainage divides are indicated.

related to the presence of hot upper mantle beneath continental crust and is an example of incipient continental rifting. Western North America is particularly interesting because it appears to have overridden the East Pacific Rise. The highest elevations, in Colorado, correspond to the boundary between a continental crust to the west which has thinned to about 20 km and overlies hot upper mantle and continental crust to the east which has been thickened to 50 km (Soller et al., 1982). The crust underlying eastern Colorado and Kansas must have acquired its abnormal thickness since the Cretaceous, when this area was a seaway. It seems likely that the deep continental crust which has been lost in the west has flowed or been underthrust to the east, resulting in the 50 km thickness measured beneath the Great Plains. The uplift in Africa occupies both the site of the East African Rift system and the Kalahari desert. Whether the relief of the Kalahari is due to thick continental crust or to hot upper mantle is not known; although crustal thickness of 36-39 km have been measured to the west in the Transvaal (Willmore et al., 1952), there are no crustal thickness measurements in this area. If the elevation of the Kalahari is due to its being underlain by hot upper mantle, it represents a pre-rift phase of uplift. If it is due to abnormally thick continental crust, it may well be crust that has been thickened by flow or underthrusting of lower continental crust west from the southern segments of the East African Rift.

It is evident that only major plate tectonic processes result in elevation of the land surface above 1 km; hence some simple generalizations can serve as guidelines in reconstructing paleotopography: 1) Continental collisions may result in uplift to 5 km; the elevation attained is a function of the combined momentum of the two colliding blocks and the time elapsed since the collision. The Alps do not appear on the 5° square average elevation map. The time of collision of southern and northern Europe is about the same as the start of the collision of India with Asia, but the momentum involved in producing the Alps was much less than the momentum that produced the Himalayas and Tibetan Plateau and the collision process was of much shorter duration in the Alps.
2) Subduction of ocean crust 15-30 m.y. old will result in uplift of 1 to 2 km and subduction of ocean crust younger than 15 m.y. may result in uplift of 2-3 km. Young oceanic lithosphere is both relatively light and thin. Newly created oceanic lithosphere loses heat rapidly at first, then more slowly as it ages and thickens. The thickening of oceanic lithosphere must be rapid at first and then decrease in rate with time. Young oceanic lithosphere may have a thickness only a fraction of typical continental crust, and can be expected to be much more flexible because of its warmth. As it is subducted beneath continental crust it is likely to be inserted between the continental crust and older colder mantle, flowing quasi horizontally until it meets a barrier, such

as the root of a mountain range, or until it has become brittle enough for fracture and shear to take place and a new underthrust to develop, resulting in stacking of sheets of oceanic lithosphere beneath the continental crust. Quantification of this special subduction process for young lithosphere has recently been attempted by Sacks (1983) for areas of western South America and Japan. 3) Incipient rifting may result in uplift of 1-2 km. Because continental collisions and subduction of hot young oceanic lithosphere are relatively rare in geologic history, but continental rifting has been quite extensive in the past this may be the most important factor in influencing detrital sediment supply over long time spans.

It is important to explore how uplift might be related to the rifting process. We assume that rifting apart of major continental blocks is intimately related to the later ongoing sea floor spreading process, so that the elevation changes associated with aging of oceanic lithosphere need to be examined in detail.

As noted above, it is generally accepted that the bathymetry of the ocean basins reflects the distribution of heat in the oceanic lithosphere (Menard, 1969; Sclater and Francheteau, 1970; Sclater, Anderson, and Bell, 1971). As the lithosphere cools by losing heat to the overlying water, it becomes more dense. It is generally stated that the depth of the sea floor can be expressed as a simple function of the age: depth below ocean ridge crest = $K \sqrt{age}$, where K = 320-350 and the age is expressed in Ma (Winterer, 1973; Kennett, 1982; Seibold and Berger, 1982). Early studies did not discriminate clearly between the depth of the sea floor and the corrected depth of oceanic crust unloaded of sediment, but later studies recognized the importance of the correction (Sclater and Detrick, 1973). Early studies also failed to take into account the accelerated cooling effect of hydrothermal circulation; this is most significant in the vicinity of the ridge crest and becomes less and less important in older crust in which the cracks have healed and which becomes overlain by progressively thicker sediment cover. Although theoretical models for loss of heat by conduction have been worked out in detail for solid lithosphere (Sclater and Tapscott, 1979) the problem of how to handle the shift from thin ocean lithosphere cooled chiefly by circulating water to older oceanic lithosphere cooled chiefly by conduction has not been solved. Hay et al. (in press) have discussed ways of obtaining empirical curves which closely describe subsidence of the unloaded surface of the oceanic crust. For the Angola and Brazil basins of the South Atlantic they derived the formula $h = 2.3644t^{0.1773}$, is the depth of unloaded basement at the given site at any age t. Because the fractional root of numbers less than 1 approaches infinity, the equation cannot be used to estimate the elevation of the Ridge at its time of origin. The term -2.3644 is the depth of the Ridge crest below present sea level predicted for crust having an age of 1 m.y. Similar empirical curves can be derived from data for the North Atlantic and part of the Pacific Ocean. The depth of the crest of the mid-ocean ridge is not everywhere 2,500 meters as is widely assumed in paleobathymetric reconstruction, but varies over extensive regions from depths as shallow as 2,000 meters to more than 3,000 meters. The cause of these differences are not known but may well have to do with cooling by hydrothermal circulation or overall lower initial temperature and/or greater initial thickness of the lithosphere probably related to hydration of upper mantle rocks. In any case, the change in average adiabatically adjusted density of the upper about 3.23 g/cc beneath the mid-ocean ridge crust to 3.30 beneath 180 m.y. old ocean crust.

Sleep (1971), Kinsman (1975), Hay (1981) and Southam and Hay (1981) have explored the changes in elevation of land surface to be expected if continental crust were to overlie upper mantle of a density similar to that beneath the mid-ocean ridge and have predicted an incremental elevation of 1.7-1.8 km of the land surface, not taking into account erosion, which would reduce this estimate, or the lack of accelerated cooling due to hydrothermal circulation, which would increase it.

Based on the assumption that all continental elevations above 1 km are associated only with either 1) anomalously thick continental crust, 2) subduction of young ocean crust, or 3) the continental rifting process, it is possible to reconstruct continental paleotopography, as shown in figures 9-12. The paleogeographic base maps used for these reconstructions are those of Barron et al. (1981). In preparing these maps, it was assumed that rifting of large continental blocks involves intrusion of hot asthenospheric material into the lower lithosphere producing a broad arch which may reach 700-1000 km across before continental separation. Formation of a sag in the center of such an uplift would occur when hot, light asthenospheric material begins to replace the lower continental crust, and brittle failure, with consequent faulting and graben formation, would occur as the hot mantle material reaches a mid-level in the continental crust. Based on general consideration of the development of the East African rift system as described by Burke and Whitman (1973) and interpreted from the Tectonic Map of Africa, and as discussed by Southam and Hay (1982), we have assumed that upward propagation of the asthenospheric intrusion proceeds through a 100 km thick lithosphere (30 km continental crust + 70 km upper mantle) at a rate of 1 km/my; and that lateral spreading of the intrusion takes place at 10 times that rate. Thus, the time required for the continental rifting process to go from inception to completion, i.e. separation of the continental fragments, is assumed to be 100 m.y. and the full width of the intrusion may be 2000 km at the time of separation although the distal parts of this resulting uplift are low and scarcely detectable from the background noise of local topographic relief. Formation of a broad depression in the center of the uplift would occur during the interval 30-20 my prior to separation; faulting and graben formation would take place over the last 20 m.y. before separation, and subsidence of the central graben below sea level would occur 5 m.y. before separation (Hay, 1981; Southam and Hay, 1981). This corresponds to what is known of the East African and Central North Atlantic rifts.

To estimate elevation of the uneroded surface at any point, the density of the asthenospheric material is assumed to be 3.23 g/cc and that of the continental crust to be 2.8 g/cc. In reality, loss of heat from the asthenospheric material to the surrounding lithosphere will cause the former to contract and become denser and the latter to expand and become lighter, but because the time scale for conduction of heat is comparable to the assumed rate of intrusion, and because the density changes compensate for each other, this complication has been ignored. The assumption of density of the asthenospheric material to be 3.23 may well be in error because this is based on analogy with the mid-ocean ridge system. The density of the asthenospheric material upwelling beneath continental crust could be significantly lighter, with resulting higher uplift; this is because the continental crust may act both as a thermal source and as a barrier to penetration of water from above.

The ancient uplifts must be projected onto an estimate of the thickness of continental crust and sediment on the continental blocks at each moment of geologic time of interest. Hay et al. (1981) estimated the average elevation of Pangaea to be 1060 m above present sea level or about 1320 meters above sea level at the time, based on the assumption that sediment presently resting on ocean crust in the continental slopes and rises and the $CaCO_3$ presently in pelagic sediment was on the continental blocks at the beginning of the Mesozoic. They neglected to add in pore space to the sedimentary mass; Southam and Hay (1982) assumed that the average sediment porosity was 22%, which adds about 500 m of thickness. The average elevation of Pangaea then becomes 1564 m above present sea level if the pore space was filled with air or 1414 m above present sea level if the pore space was filled with water and the elevation isostatically adjusted. Sea level at the time would have been about 60 m below the present level if the pore space was air filled, or 225 m below present sea level if the pore space was water filled. The average elevation of Pangaea above sea level at the time is essentially independent of whether the pore space in the sediments on the continental blocks was filled with air or water,

these differences being compensated isostatically to yield a figure of 1624 m
for the dry pore space model and 1639 m for the wet pore space model.

To construct the paleotopographic maps Hay (1984) assumed a constant rate of
loss of sediment off the continental blocks. The thermal elevation increments
predicted for each 5° square were added to the predicted average elevation of
the continents for each time. If the average elevation of the continents was
predicted to be more than 1 km, the coastal areas which were not active margins
or young passive margins were assumed to be less than 1 km. To compensate for
this additional elevation increments were added to the remainder of the
continental area; with the coarse 1 km vertical resolution, these are minor
adjustments. The average background elevation argument is based on
consideration of the overall off-loading of sediment from the continental
blocks onto oceanic crust and neglects the short term variations in sediment
supply to the deep sea (Southam and Hay, 1982). Adding elevation increments
onto this background should increase the average elevation of the continents
for any given time. However, the shorelines were changing as sea level rose to
+300m (Sleep, 1976) in the mid Cretaceous and has since fallen back to its
present level. The effects of higher sea levels is to reduce the average
elevation of the continents. Figure 8 shows the combined effects of sea level
change resulting from sea floor spreading processes and two models for average
continental elevation through time. The upper model (A) represents the change
in average elevation from 220 Ma to present assuming constant rate loss
sediment, ignoring thermal effects. The lower model (B), which was used in
constructing the paleotopographic maps, presented here, was determined by
assuming a constant rate of loss of sediment from 200 Ma to a hypothetical
present day average elevation of about 200m derived by subtracting all of the
present areas with "anomalous" elevations due to plate tectonic processes,
active only during the past 30 m.y. (i.e. continental collisions and large
scale subduction of young oceanic lithosphere) from the present day average
elevation. The present anomalous elevations are readily seen on the plots of
hypsographic curves of individual continents (Hay and Southam, 1977, fig. 9;
Hay et al., 1981, fig. 3; Harrison et al., 1981, 1983), from the global
hypsographic curve.

5 GENERAL GEOGRAPHIC AND TOPOGRAPHIC CHANGES OVER THE PAST 160 my

Figures 9-12 are 5° square average elevation maps with a 1 km contour
interval, for 40, 80, 120 and 160 m.y. Compared with the map of the Recent
(fig. 7), these show the general trends of geographic and topographic change
over the past 160 m.y. The shelf seas and dominant shelf sediment type are also
shown and will be discussed in the next section.

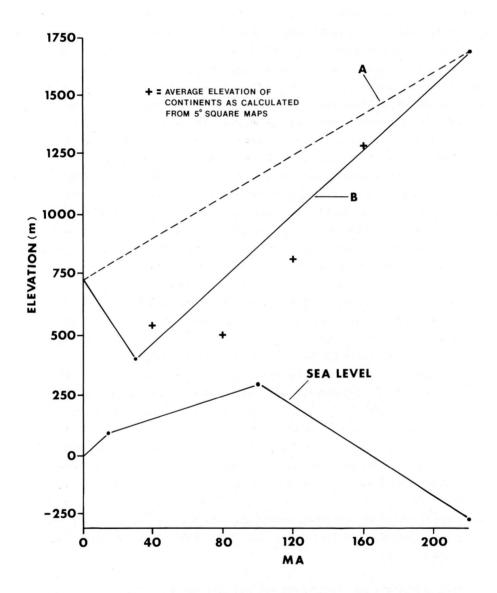

Fig. 8. Decrease in average elevation of the continents and change in sea level resulting from development of the mid-ocean ridge system; both with reference to present sea level. A is constant rate of decrease to 30 Ma, then increase in average elevation resulting from collision of the Indian subcontinent with Asia and overriding of the Pacific spreading center by western North America. Crosses are average elevations of the continents determined from the 5° square average elevation maps (figures 9-12).

23

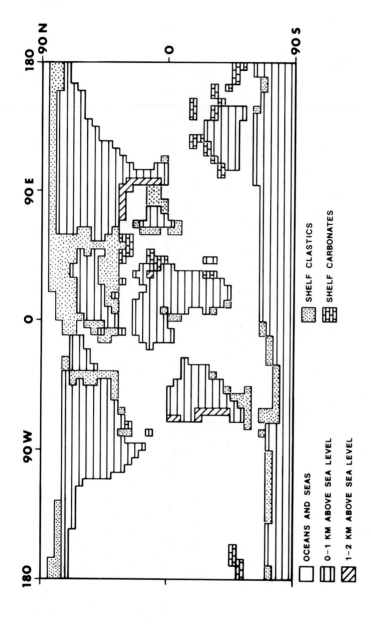

Fig. 9. Predicted average elevation and drainage divides, and observed sediment types (after Habicht, 1979, and Termier and Termier, 1952) for 5° squares at 40 Ma.

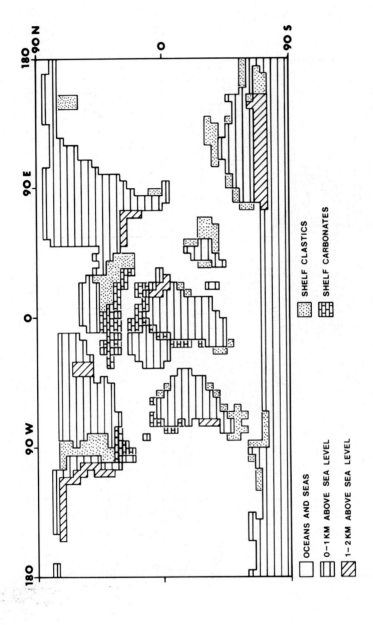

Fig. 10. Predicted average elevation and drainage divides, and observed sediment types (after Habicht, 1979, and Termier and Termier, 1952) for 5° squares at 80 Ma.

25

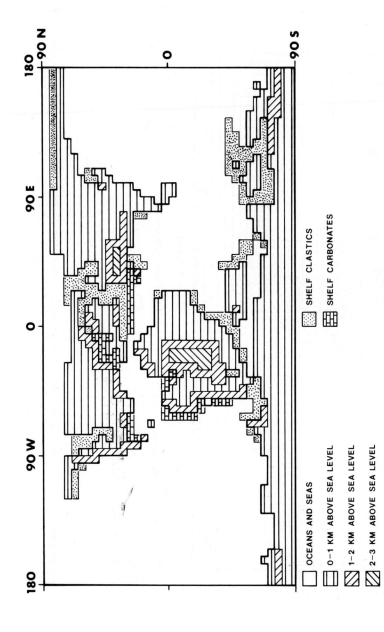

Fig. 11. Predicted average elevation and drainage divides, and observed sediment types (after Habicht, 1979, and Termier and Termier, 1952) for 5° squares at 120 Ma.

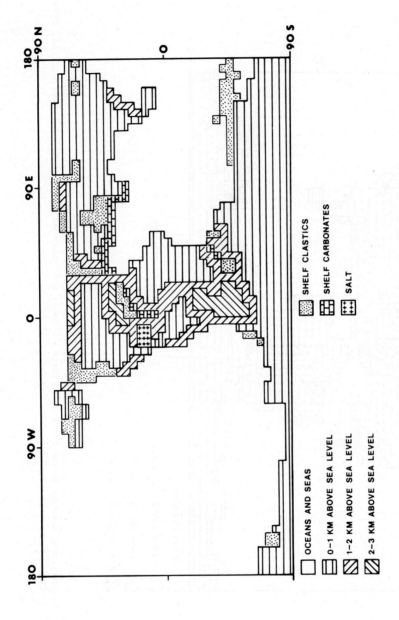

Fig. 12. Predicted average elevation and drainage divides, and observed sediment types (after Habicht, 1979, and Termier and Termier, 1952) for 5° squares at 1600 Ma.

It is evident that for much of the geologic past, and particularly the Mesozoic, the source areas for detrital sediment were uplifts on the sites of continental rifting. These uplifts exposed granitic and granodioritic rocks with free quartz to weathering and erosion. As a result, the sediments derived from these sources should contain sand bodies with high porosity and permeability. It is important to realize that rivers could carry this material long distances, across entire continents resulting from fragmentation of Pangaea.

6 SEDIMENTATION OF TERRIGENOUS DETRITUS AND CARBONATE ON CONTINENTAL MARGINS

The observed dominant shelf and epeiric sea sediment types are also shown on the maps of present day and predicted ancient topography. Information on rates of sedimentation is inadequate and is not indicated.

On the 5° square map of the Recent (fig. 7) carbonate dominated shelves extend to 30° N and 35° S. They are typically far from uplifts; a notable exception is the young ocean situation of the Red Sea, where drainage is directed away from the rifted margin and carbonate sedimentation dominates the narrow shelves.

The map for 40 Ma (fig. 9) shows carbonate dominated shelves extending to 35° N and 55° S. This greater latitudinal extent of areas of carbonate deposition is a result of the warm equable climates of the Eocene. The extensive epeiric and shelf seas of North America and Asia are dominated by terrigenous sediment, although in fact the sediments contain large proportions of carbonate shell material in many areas.

At 80 Ma (fig. 10) carbonates dominate shelf areas from 45° N to 30° S, being particularly prominent along the shelves of the Tethys seaway. Distribution of carbonate dominated shelf seas in the Campanian does not appear to cover as great a spread of latitudes as in the Eocene. This may reflect slightly lower global temperatures during the late Cretaceous, as suggested by oxygen isotope data (Savin, 1977), although the climates were equable and equator to pole temperature gradients much less than at present. Terrigenous detrital sediments dominate in higher latitudes. In the western interior seaway of North America, detrital sediments accumulated to the east of the western marginal uplift. Extensive areas of detrital sediments on the shelves of southern South America and Australia may also be related to uplift along the western margin of southern South America and on the site of the separation of Australia and Antarctica.

At 120 Ma (fig. 11) conditions were generally similar to those of the late Cretaceous. Carbonate dominated shelves extended from 50° N to 45° S, again reflecting the more equable climate of the time. Carbonate dominated shelves again border the Tethys. Detrital sedimentation, derived from western sources, dominated the North American western interior seaway. Epeiric seas in Asia appear to be dominated by terrigenous detrital sediments shed from the south-central Asian uplifted area. Detrital sediments along the eastern margin of Africa were probably derived from the uplift on the site of the future South Atlantic. Detrital sediments of the epeiric seas and shelves of Australia were probably derived from the uplift on the site of separation of the Campbell Plateau/New Zealand from Australia.

At 160 Ma (fig. 12) the shelf sea areas were dominated by carbonates from 45° N to 40° S, especially along the northern margin of the Tethys and the eastern margin of the young Central North Atlantic. The western shelves of the young Central North Atlantic are dominated by terrigenous detritus. Sedimentation on the margins of the Central North Atlantic was dominated by the climatologic effects produced by the uplifted margins in the northern belt of Easterly Trade winds, as discussed by Hay et al. (1982). The extensive salt deposition in the Gulf of Mexico was a result of extreme evaporative conditions resulting from the enclosure of the basin by uplifts to the north, east and south (Hay et al., 1982).

Although these maps are necessarily crude, they do begin to offer insight into the nature of global tectonic control of the shelf and epeiric sea sedimentation patterns. More specifically the patterns on active and passive continental margins are markedly different. Sedimentation on active margins is dominated by the active tectonic setting of subduction process and carbonates are virtually excluded. Sedimentation on passive margins is controlled by subsidence processes resulting from cooling of the lithosphere and the special oceanographic conditions prevailing in shelf seas. Detrital sediments are derived from the point sources of river mouths, whereas carbonate is potentially available everywhere as it is dissolved in the ocean reservoir.

It will be of great importance to add data on sedimentation rates to the maps so that the response of the sedimentation regime to uplift rates can be interpreted. At present this is difficult to do because of the limited amount of information presently available on the extent, thickness and volumes of older strata along the continental margins.

In reality the availability of carbonate for deposition over extensive areas appears to be related to the equator to pole temperature gradient. Although organisms are able to extract carbonate from sea water as shelf materials everywhere in the ocean and extraction may even produce shelly deposits in the polar regions, carbonate sedimentation dominates only where waters are relatively warm, presently between the latitudes of 30° N and 35° S. Figures 9-12 indicate greater latitudinal extent of large scale carbonate deposition in the past. A prerequisite for deposition of carbonate in shallow water is the absence of a source of terrigenous detritus, which can be supplied much more rapidly than carbonate. The region around river mouths are overloaded with detritus, but once delivered to the shelf sea, redistribution processes become increasingly less effective at spreading the terrigenous detrital sediment further and further away from the point source.

Once carbonate becomes a dominant sediment in a shelf area, it tends to exclude terrigenous detrital sediment by building up the sea floor to very shallow depths. Figure 13 is a diagram of the depth of the shelf break of the eastern US versus latitude taken from the AAPG maps of the bathymetry of the eastern continental margin of the U.S. (Holland, 1970); a similar diagram has been presented by Uchupi (1968). Off southern Florida, where the shelf is dominated by carbonates, the shelf break is at a depth of only 10 m. Off New England, the shelf break is typically at depths of 120-180 m. The reason for this effect is the cementation of carbonate material which occurs as sea level drops and fresh water can percolate through the deposits. The cemented carbonates are not subject to mechanical erosion as are terrigenous detrital shelf sediments. During the Pleistocene, the shelf regions dominated by terrigenous detritus have become adjusted to the glacial low stands of sea level. The river valleys cut during glacial low sea levels are flooded and form estuaries, and relict deposits mark the relatively deep shelf break. The dominated shelves reflect the interglacial high stands of sea level with a very shallow shelf break. Because the carbonate shelves build up during high stands of sea level and cannot be eroded during low stands, they tend to exclude terrigenous detritus.

Rivers produce shelf sedimentation conditions which tend to exclude carbonates; carbonate buildups, being topographic highs tend to exclude terrigenous detritus. If rivers enter a carbonate area in the present tropics or subtropics, they may cut a channel and bypass their terrigenous load across the shelf into the deep sea. Strong facies contrasts are to be expected as a result of the tendency of the two sediment types to exclude each other.

7 SUMMARY AND CONCLUSIONS

The Deep Sea Drilling Project has provided a data base which permits us to begin to understand how plate tectonics controls the global distribution of detrital and carbonate sediments, but much more knowledge from the continental margins is needed to complete the picture.

Terrigenous detritus is delivered to the sea chiefly by rivers, and hence has point sources along the coastlines. Carbonate is carried in solution to the sea and is potentially available everywhere by drawing on the ocean reservoir.

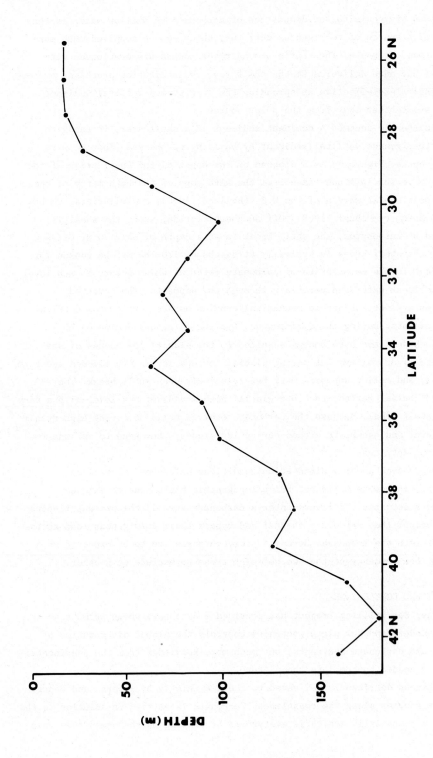

Fig. 13. Depth of the shelf break off the east coast of the U.S. as a function of latitude.

Terrigenous detrital sediments are spread over the shelves from the point sources and having higher rates of local supply, tend to exclude carbonates. Carbonates are deposited where waters are warm and there is no major source of detrital sediment. Carbonate buildups tend to exclude detrital sediment by virtue of their relief.

The supply of terrigenous detritus by rivers is a function of the elevation of the drainage basin, and to a lesser extent, the local climatology. At present, the Himalayan-Tibetan drainage system supplies about one fifth of all the terrigenous detritus delivered to the sea. Hence, knowledge of paleotopography of the continents is the key to predicting the detrital loads of rivers of the past.

Paleotopography can be reconstructed using information on the average elevation of the continents in the past, derived from calculations of the volumes of sediment offloaded from the continental blocks since the breakup of Pangaea, and by assuming that large elevated areas are produced only by continental collision, subduction of young ocean lithosphere, and during the rifting process.

Maps of 5° averages showing reconstructed paleotopography and observed dominant shelf sediment types, constructed for 40, 80, 120, and 160 Ma compare favorably with a map of the Recent and appear to be internally consistent. This suggests that topographic relief is required as a source of terrigenous detritus, and that the area available for carbonate deposition is limited by the input of terrigenous detritus as well as changing equator to pole temperature gradients.

8 ACKNOWLEDGEMENTS

During the course of this work we have benefited from discussions with T. A. Davies, C. G. A. Harrison, J. R. Southam, E. J. Barron and numerous others, but the conclusions presented here are the responsibility of the authors. We are especially indebted to Paul Ewbank and Kim Miskell at the Rosenstiel School of Marine Atmospheric Sciences, University of Miami, who wrote and ran the computer program to calculate the hypsography of the drainage basins delineated by M. Rosol. This work has been supported by grant NSF-OCE-8218914 from the Marine Geology Program of the National Science Foundation.

9 REFERENCES

Barron, E.J., Harrison, C.G.A., Sloan, J.L. II and Hay, W.W., 1981.
Paleogeography, 180 million years ago to the present. Eclogae Geol. Helvet.,
74: 443-470.
Berger, W.H., 1979. The impact of Deep-Sea Drilling on paleoceanography. In:
M. Talwani, W. Hay and W.B.F. Ryan (Editors), Deep Drilling Results in the
Atlantic Ocean: Continental Margins and Paleoenvironment. Amer. Geophys.
Union Maurice Ewing Series, 3: 297-314.
Burke, K. and Whiteman, A.J., 1973. Uplift, rifting and the break-up of
Africa. In: D.H. Tarling and S.K. Runcorn (Editors), Implication of
Continental Drift to the Earth Sciences. Academic Press, London,
pp. 735-755.
Cande, S.C. and Mutter, J.C., 1982. A revised identification of the oldest
sea-floor spreading anomalies between Australia and Antarctica. Earth
Planet. Sci. Letters, 58: 151-160.
Davies, T.A. and Worsley, T.R., 1981. Paleoenvironmental implications of
oceanic carbonate sedimentation rates. Soc. Econ. Paleont. Mineral. Spec.
Pub., 32: 169-179.
Davies, T.A., Luyendyk, B.P., Kidd, R.B. and Weser, O.E., 1975. Unconformities
in the sediments of the Indian Ocean. Nature, 253: 15-19.
Davies, T.A., Hay, W.W., Southam, J.R. and Worsley, T.R., 1977. Estimates of
Cenozoic oceanic sedimentation rates. Science, 197: 53-55.
Garrels, R.M. and Mackenzie, F.T., 1971. Evolution of Sedimentary Rocks.
W.W. Norton Co., New York, 397 pp.
Habicht, J.K.A., 1979. Paleoclimate, Paleomagnetism and Continental Drift.
Amer. Assoc. Petrol. Geol., Studies in Geology, 9: 31.
Harrison, C.G.A., Brass, G.W., Saltzman, E., Sloan, J. II, Southam, J. and
Whitman, J.M., 1981. Sea level variations, global sedimentation rates, and
the hypsographic curve. Earth Planet. Sci. Letters, 54: 1-16.
Harrison, C.G.A., Miskell, K.J., Brass, G.W., Saltzman, E.S. and
Sloan, J.L. II, 1983. Continental hypsography. Tectonics, 2: 357-377.
Hay, W.W., 1981. Sedimentological and geochemical trends resulting from the
breakup of Pangaea. Oceanologica Acta, 4, Suppl.: 135-147.
Hay, W.W., 1983. Significance of runoff to paleoceanographic conditions during
the Mesozoic and clues to locate sites of ancient river inputs: V Joint
Oceanographic Assembly. Proc. Dept. Fisheries and Oceans, Ottawa: 9-17.
Hay, W.W. The breakup of Pangaea: Climatic, erosional and sedimentological
effects. Proc. Internat. Geol. Congr., Moscow, 1984.
Hay, W.W. and Southam, J.R., 1976. Calcareous plankton and loss of CaO from
the continents. Geol. Soc. Am. Abstracts with Programs, 7: 1105.
Hay, W.W. and Southam, J.R., 1977. Modulation of marine sedimentation by the
continental shelves. In: N.R. Andersen and A. Malahoff (Editors), The Fate
of Fossil Fuel CO_2 in the oceans. Plenum Press, New York, pp. 569-604.
Hay, W.W., Barron, E.J., Sloan, J.L. II and Southam, J.R., 1981. Continental
drift and the global pattern of sedimentation. Geologische Rundschau, 70:
302-315.
Hay, W.W., Barron, E.J., Behensky, J.F. Jr. and Sloan, J.L. II, 1982.
Triassic-Liassic paleoclimatology and sedimentation in proto-Atlantic rifts.
Paleogeogr., Paleoclimatol., Paleoecol., 40: 13-30.
Hay, W.W., Thompson, S.L. and Barron, E.J., 1982. Large volume, low latitude,
evaporite accumulation in the Gulf of Mexico rift. Geol. Soc. Amer. Abstr.
in Prog., 14: 511.
Hay, W.W., Sibuet, J.-C., Barron, E.J., Boyce, R.E., Brassell, S., Dean, W.E.,
Huc, A.Y., Keating, B.H., McNulty, C.L., Meyers, P.A., Nohara, M.,
Schallreuter, R.E., Steinmetz, J.C., Stow, D. and Stradner, H. (in press).
Site 530. Initial Reports of the Deep Sea Drilling Project. U.S. Government
Printing Office, Washington, 75.

Holland, H.D., 1978. The Chemistry of the Atmosphere and Oceans. John Wiley and Sons, New York, 351 pp.

Holland, H.D., 1982. River transport to the oceans. In: C. Emiliani (Editor), The Oceanic Lithosphere, The Sea, 7. Wiley, New York, pp. 763-800.

Holland, W.C., 1970. Bathymetric Maps, Eastern Continental Margin: U.S.A. Amer. Assoc. Petrol. Geol., 3 sheets.

Kennett, J.P., 1982. Marine Geology. Prentice-Hall, Englewood Cliffs, N.J., xv + 813 pp.

Kinsman, D.J.J., 1975. Rift valley basins and sedimentary history of trailing continental margins. In: A.G. Fischer and S. Judson (Editors), Petroleum and Global Tectonics. Princeton University Press, Princeton, N.J., pp. 83-126.

Lerman, A., 1982. Controls on river water composition and the mass balance of river systems. In: River Inputs to Ocean Systems. United Nations, New York, pp. 1-4.

Lisitzin, A.P., 1974. Osadkoobrazovanie v Okeanach. Izdatelstvo "Nauka," 438 pp.

Livaccri, R.F., Burke, K. and Sengor, A.M.C., 1981. Was the Laramide Orogeny related to subduction of an oceanic plateau? Nature, 289: 276-278.

Menard, H.W., 1969. Elevation and subsidence of oceanic crust: Earth Planet Sci. Lett., 6: 275-284.

Milliman, J.D., 1974. Marine carbonates. In: Milliman, J.D., Mueller, G., and Foerster, U. (eds.), Recent Sedimentary Carbonates. Springer Verlag, Heidelberg, pt. 1, xv + 375 pp.

Moore, T.C., 1972. DSDP: successes, failures, proposals. Geotimes 17: 27-31.

Moore, T.C., and Heath, G.R., 1977. Survival of deep sea sedimentary sections: Earth Planet. Sci. Letters, 37: 71-80.

Poldervaart, A., 1955. Chemistry of the earth's crust. In: Poldervaart, A. (ed.), Crust of the Earth. Geol. Soc. Amer., Spec. Pap., 62: 119-144.

Prospero, J.M., 1982. Eolian transport to the world ocean. In: Emiliani, C. (ed.), The Oceanic Lithosphere, The Sea, v.7. Wiley, New York, p. 801-874.

Rona, P.A., 1973. Relations between rates of sediment accumulation on continental shelves, sea floor spreading, and eustasy inferred from the Central North Atlantic: Geol. Sci. Am. Bull., 84: 2852-2871.

Ronov, A. B., 1959. On the post-Precambrian geochemical history of the atmosphere and hydrosphere. Geochemistry, 5: 493-506.

Ronov, A.B., 1964. Common tendencies in the chemical evolution of the earth's crust, ocean and atmosphere. Geochemistry, 8: 715-743.

Ronov, A.B., 1982. The earth's sedimentary shell (quantitative patterns of its structures, compositions, and evolution). The 20th V.I. Vernadskiy Lecture, March 12, 1978. Internat. Geol. Rev., 24: 1365-1388.

Sacks, I.S., 1983. The subduction of young lithosphere. J. Geophys. Res., 88: 3355-3366.

Salvador, A. and Green, A.R., 1981. Opening of the Caribbean Tethys. 26th Congr. Geol. Internat., Colloque C5, Geologie des chaines alpines issues de la Tethys. Bur. Rech. Geol. et Min., Mem. 115: 224-229.

Savin, S.M., 1977. The history of the earth's surface temperature during the past 100 million years. Ann. Rev. Earth Planet. Sci., 5: 319-355.

Sclater, J.G. and Detrick, R., 1973. Elevation of Midocean ridges and the basement age of JOIDES Deep-Sea Drilling Sites. Geol. Soc. Amer. Bull, 84: 1547-1154.

Sclater, J.G. and Francheteau, J., 1970. The implications of terrestrial heat-flow observations on current tectonic and geochemical models of the crust and upper mantle of the earth. Geophys. J. Roy. Ast. Soc., 20: 509-542.

Sclater, J.G. and Tapscott, C., 1979. The history of the Atlantic. Scientific American, 204: 156-174.

34

Sclater, J.G., Anderson, R.N. and Bell, M.L., 1971. Elevation of ridges and evolution of the Central Pacific. J. Geophys. Res., 76: 7888-7915.

Seibold, E. and Berger, W.H., 1982. The Sea Floor: An Introduction to Marine Geology. Springer-Verlag, Heidelberg, vii + 288 pp.

Sibley, D.F. and Vogel, T.A., 1976. Chemical mass balance of the earth's crust: The calcium dilemma and the role of pelagic sediments. Science, 192: 551-553.

Sibley, D.F. and Wilband, J.T., 1977. Chemical balance of the earth's crust. Geochim. Cosmochim. Acta, 41: 545-554.

Sleep, N.H., 1971. Thermal effects of the formation of Atlantic continental margins by continental breakup. Geophys. J. Roy. Ast. Soc., 24: 325-350.

Sleep, N.H., 1976. Platform subsidence mechanism and "eustatic" sea level changes. Tectonophysics, 36: 45-56.

Soller, D.R., Ray, R.D. and Brown, R.D., 1981. A new global crustal thickness map. Tectonics, 1: 125-139.

Southam, J.R. and Hay, W.W., 1977. Time scales and dynamic models of deep-sea sedimentation. J. Geophys. Res., 82: 3825-3842.

Southam, J.R. and Hay, W.W., 1981. Global sedimentary mass balance and sea level changes. In: C. Emiliani (Editor), The Oceanic Lithosphere, The Sea, 7. Wiley, New York, pp. 1617-1684.

Termier, H. and Termier, G. Histoire geologique de la Biosphere. Masson et Cie., Paris, 721 pp.

Uchupi, E., 1968. Atlantic continental shelf and slope of the United States Physiography. U.S. Geol. Surv. Prof. Pap. 529-C, 30 pp.

Vail, R.R., Mitchum, R.M. Jr., and Thompson, S. III, 1977. Seismic stratigraphy and global changes of seal level, part 4: Global cycles of relative changes of sea level. In: C.E. Payton (Editor), Seismic Stratigraphy--Applications to Hydrocarbon Explorations. Am. Assoc. Petrol. Geol. Mem. 26: 83-97.

Weissel, J.K., Hayes, D.E. and Herron, E.M., 1977. Plate tectonics synthesis; the displacements between Australia, New Zealand, and Antarctica since the Late Cretaceous. Marine Geology, 25: 231-277.

Whitman, J.M. and Davies, T.A., 1979. Cenozoic oceanic sedimentation rates: how good are the data? Marine Geology, 30: 269-284.

Willmore, P.L., Hales, A.L. and Gane, P.G., 1952. Seismic investigation of crustal structure in western Transvaal. Seismol. Soc. Amer. Bull, 42: 53-80.

Winterer, E.L., 1973. Sedimentary facies and plate tectonics of equatorial Pacific. Am. Assoc. Petrol. Geol. Bull., 57: 265-282.

Worsley, T.R. and Davies, T.A., 1972a. Sea level fluctuations and deep-sea sedimentation rates. Science, 203: 455-456.

Worsley, T.R. and Davies, T.A., 1979b. Cenozoic sedimentation in the Pacific Ocean: steps toward a quantitative evaluation. Jour. Sed. Petrology, 49: 1131-1146.

Chapter 2

SEDIMENTARY MODELS OF SILICICLASTIC DEPOSITS AND CORAL REEFS INTERRELATION

G. SANTISTEBAN AND C. TABERNER Departamento de Geologia, Fctad. de Ciencias Biologicas, Universidad de Valencia, Dr. Moliner, 50, Burjassot (Valenoia), Spain; Departamento de Petrologia, Fctad. de Geologia. Universidad de Barcelona, Gran Via, 585, Barcelona 7. Spain.

ABSTRACT
    In the geological record of two differently positioned basins near the Spanish Mediterranean coast (the Biocene Fortuna Basin and the Eocene Catalan Basin) there exists a great number of reefs physically related to siliciclastic deposits. These reefs developed in different environments (delta, fan-delta, beach and tidal environments) coexisting with an active terrigenous sedimentation. In six cases of those analyzed, the sedimentary bodies of siliclastics (bars, lobes, channels) controlled the geometry and positioning of the reefs. In one of them, the reef conditioned the types of facies and positioning of the terrigenous sediments.

1 INTRODUCTION

    The existence of reef buildups between siliciclastic sediments has not been emphasized until recently. Usually in these cases, siliciclastics and carbonates have been studied separately; perhaps it is due to the high level of specialization attained in sedimentology and also to the different methods of study applied in both kinds of deposits.

    A lot of information about depositional environments in a mixed platform can be supplied by the isolated study of siliciclastics or of carbonates; however, analyzing both of them as a whole allows us to obtain more documentation. This can be especially important when there exists evidence that there was dynamic control of processes and products of the siliciclastic environments by the carbonatic ones, or vice versa.

    Some examples of coral reefs and other kinds of carbonate buildups of organic origin, related to environments in which active siliciclastic sedimentation is required, have been described in the geological literature (Brown, 1969; Brown et al., 1973, Brown and Fisher, 1977; Dabrio, 1975; Sanristeban and Taberner, 1977, 1979, 1980; Hayward, 1982; Choi and Ginsburg, 1982; Choi and Holmes, 1982; Reid and Tempelman-Kluit, 1982). In most of these examples cited, organic buildups were developed on delta or fan-delta lobes. There are various sedimentary models in which recent or subactual coral reef growth on active delta or fan-delta systems has been described (Gvirtzman and Buchbinder, 1978; Guilcher, 1979; Wescott and Ethridge, 1980; Hayward, 1982; Friedman, 1982).

The existence of an active clastic (terrigenous or carbonatic) sedimentation in coastal systems with coral reefs, brings about four main factors that cause a negative effect on the development of corals: 1) water turbulence, 2) sedimentation of particles, 3) supply of fresh water, if currents come directly from the continent, and 4) mechanical erosion due to high-energy sporadic currents. The prolonged action of any of these negative agents on a reef system is enough to cause the death of their components. Because of this, coeval reef growth and active clastic sedimentation have been considered incompatible.

However, in the fossil record it is quite common to find coral reefs physically related to terrigenous sediments. These reefs show proof that allows us to consider that they were dynamically controlled by the sedimentary processes of delta or fan-delta systems. This kind of relationship could be explained once the pattern of incidence of the four negative factors on the reef communities mentioned above, has been determined. Because of this, it is necessary to consider the relationships between terrigenous sedimentation and the development of organic buildup attending the true interval of time in which it is developing. So, for example, a delta and a fan-delta system have different sedimentary rhythms and the relationship between siliciclastics and reefs has to be considered in a different light.

Through this paper we describe different examples of the interrelation between siliciclastics and reef carbonates that have been found in two Tertiary sedimentary basins placed close to the Spanish Mediterranean coast (Fig. 1). Most of these examples have been studied in the Eocene Catalan Basin (NE Iberian Peninsula, Fig. 1B), while some of them are also represented in the Upper Miocene (Tortonian) sediments of the Fortuna Basin (Fig. 1A). These examples comprise a wide spectrum of relationships between reefs and terrigenous sediments developed in some sedimentary environments (fan - delta, delta, tidal, beaches, etc.). In these environments reefs develop on sedimentary bodies having a lobe shape or on the bed of a channel.

The relationships between reefs and siliciclastic bodies that we have investigated are: 1) morphologic, and 2) dynamic. The study of the relationships between siliciclastics and reefs could help in the progress for the knowledge of mixed facies.

2 CORAL REEFS PLACED ON FAN-DELTA SYSTEMS

The existence of reefs placed on fan-delta systems has been recognized in both studied sedimentary basins. However, the examples better represented and documented are the ones studied in the Tortonian deposits of the Fortuna Basin (Fig. 1A). In this basin there are three places where deposits corresponding to

37

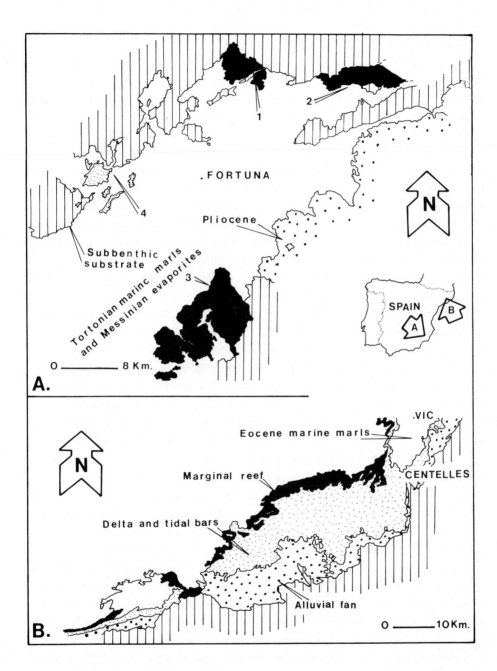

Fig. 1. Geological schematic maps of the two basins exhibiting the examples
described in this work. A) Geological map of the Fortuna Basin (1, 2, and
3, reefs related with fan-delta deposits, 4, delta). B) Partial
geological map of the Eocene Catalan Basin.

three different fan delta systems have been preserved (Fig. 1A). External
configuration and original facies relationships display excellent preservation
in these deposits.

The largest fan-delta is located in the southern margin of the basin. Its
surface is 52.5 km^2. Each one of the other two fan-delta systems is ten times
smaller. The fan-delta systems consist of accumulations of conglomerates placed
between the sub-benthic substratum and marine marls. External configuration
shows sub-horizontal platforms bounded frontally by an original depositional
talus (Fig. 2). The mean difference of height between two extreme deposits is
about 200 meters. From an aerial view these fan-delta systems show a fan-shape
with a scalloped rim. In each one of these fan-delta systems minor lobe shaped
units arranged in a divergent pattern can also be distinguished (Fig. 2).

The studied fan-delta systems in both basins show basically the same kind
of facies. These facies may be grouped in two associations which represent each
one of the most significant sub-environments that can be identified in these
systems. These facies associations are: 1) alluvial facies association, and 2)
depositional lobes facies association (Fig. 2).

The depositional lobes facies association is the facies association that
specifically concerns us, with regard to the relationship between
siliciclastics and reef carbonate deposits. This association consists of
conglomerates and sandstones which are distributed to form sedimentary bodies
with lobe configuration (Fig. 2). The association shows evidence of having been
deposited in a marine environment. The marine origin of these sediments is
proved by: 1) being interfingered with marine facies, 2) acting as support of
reef buildup installation, and 3) showing borings in most of the pebbles.

Length of these bodies (in proximal - distal direction) may oscillate
between 100 and 600 m. The mean height usually varies between 15 and 45 m.
These lobes are characterized by their convex - concave large scale
cross-bedding (Fig. 3). The cross-bedding surfaces reflect in their dimension
and in their gradient, the morphological characteristics of the upper part of
the fan-delta front in each one of the lobes accretionary phases.

Three different facies may be distinguished in this association. Each one
of the three facies corresponds to each one of the morphological zones in which
a depositional lobe can be subdivided as follows: proximal, intermediate and
distal. The proximal facies is usually present in the highest part of a lobe.
Conglomerates and microconglomerate deposits arranged in amalgamated braided
channels constitute this facies. The channels show an inner structure (fining
and thinning upwards) similar to the ones of alluvial origin. The channels
differ from the alluvial ones because the former are not included in red
mudstones and because a great percentage of pebbles infilling the channels show

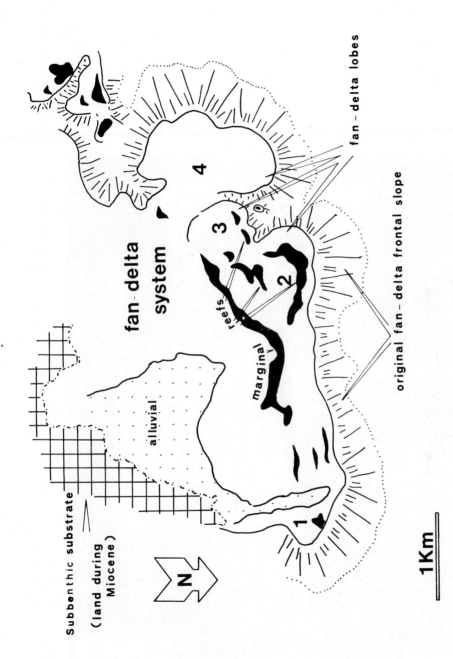

Fig. 2. Cartography of the "El Montanal" fan-delta in the Miocene Fortuna Basin.

40

Fig. 3. Reef (R) located between two fan-delta lobes. The reef developed on the frontal slope angle breaking point of the inferior lobe (L1). Later it was partially eroded and covered by the deposits of the second lobe (L2).

borings. The fact that some of these channels totally or partially cut through thick reef build-ups suggests a high erosional capacity of the currents that originated the channels.

The facies that represents the intermediate term corresponds to the deposits which originated along the frontal talus of the lobe. This facies consists of conglomerate and sandstones layers. Conglomerates are more common in the deposits of the upper part of this talus, while sandstones are gradually more important towards the lower part. These deposits show a pronounced large scale cross-bedding (20°) suggesting they originated along high depositional slopes (Fig. 3).

Sandstones, and occasionally conglomerates, constitute the distal term; they usually consist of 15-20 cm thick sandstone beds. The sandstone layers are intercalated with sandy mudstones. Normally the sandstones are strongly bioturbated, so that it is difficult to observe their inner structures. When there is no bioturbation we can state that the sandstone layers are constituted by two superimposed intervals. The lower one of these intervals shows a fining sequence or parallel lamination. The upper term exhibits climbing ripples.

The three terms (lower, middle and upper) pass gradually, horizontally and vertically, one into the other. Progradation of the front of the fan-delta system seems to be the cause of the facies superimposition.

These three facies are always arranged forming depositional bodies, showing a lobe configuration from an aerial view. These lobes are the product of the accumulation of sediment transported by a braided river system and deposited at its mouth.

2.1 Reefs Developed on Fan delta Depositional Lobes

The reefs developed on fan-delta systems are closely related to the deposits corresponding to the depositional lobes. These reefs have an elongated shape; usually they are arranged parallel to the supposed coast line. The lateral continuity is variable; it may oscillate between 10 m and 5km (Fig. 2).

In general the smallest reefs (less than 300 m length) have an arc-like configuration from an aerial view, due to their adaptation to a solitary fan-delta lobe. The bigger reefs are usually scalloped due to the fact they are bordering a larger unit constituted by an ensemble of depositional lobes (Fig. 2).

Through cartography we can appreciate that there are various arc-like reefs parallel to each other (Fig. 2). These arc-like reefs correspond to coral buildups developed during different moments of the fan-delta system evolution. Reefs are always located on the depositional lobes, at the frontal-slope angle breaking point (Figs. 3, 4, and 5). In a transverse section they are not quite larger than 200 m length. In practice, they do not present well defined lagoon areas. Talus-reef deposits are volumetrically important, in relation with the ones of the constructed area (Figs. 4 and 5). The distribution of the talus is

Fig. 4. Reef adapted to the frontal slope of a fan-delta lobe. The reef build-up (R) is situated on the break of the angle of the frontal slope. The reef-talus deposits (T) are accumulated in the depression existing between two coalescent lobes (L).

irregular due to the fact that the detritus of the reefs are preferentially accumulated in the depressions of the foundations (Fig. 4). These depressions correspond to the areas comprised between coalescent fan-delta lobes.

Reefs seem clearly morphologically adapted to the substratum constituted by the fan-delta depositional lobes. Installation of these reefs must have been accomplished during a dormant stage of the channel system that furnished the lobes; however, field evidence shows that the fan-delta was still active and that the lobes, on which reefs were developed, were not completely abandoned. The existence of various parallel arc-like reefs indicates that the lobes continued being active during the development of reefs on their fore-side. In some cases the reefs are found separated from each other and completely

43

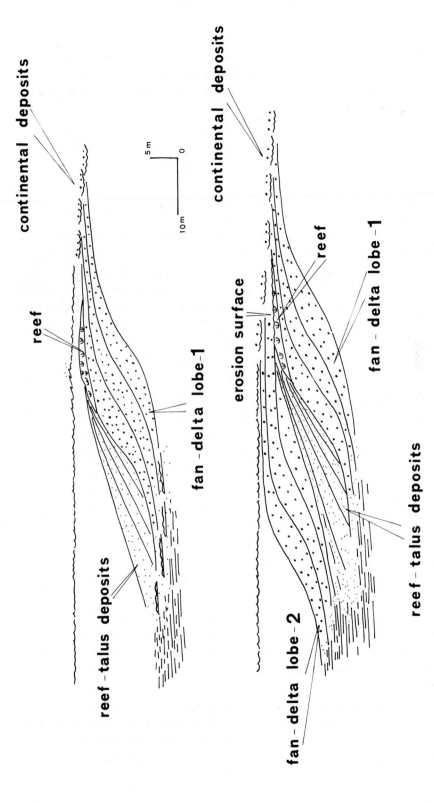

Fig. 5. Sketch of the development of a reef located between two fan-delta lobes.

isolated among conglomerates. In other cases, intercalated between the
reef-talus deposits are conglomerate wedges. This could be due to the fact that
the advance of the fan-delta lobe did not completely destroy the reef, and the
buildup could later recover its coralline development.

The shape and distribution of the reefs associated with the fan-delta
lobes enable us to establish a detailed reconstruction of the morphological
features and sequential development of the fan-delta. So, for example, in the
cartography of the fan-delta of "El Montanal" (Fig. 2) in the Fortuna Basin, we
can distinguish three different fan-delta units, keeping in mind the
development of the reefs. Each one of these units corresponds to a group of
deltaic lobes developed at the same stage. The sedimentation area of the
fan-delta system migrated through time from east to west. Unit 1 (Fig. 2)
corresponds to the most ancient deposits and Unit 3 (the same figure) to the
most recent. In Unit 1, the reefs are small and not well developed. In Unit 2,
there is good development of various arc-like fringing reefs intercalated with
terrigenous deposits, as a consequence of the fan-delta progradation. The
existence of some arc-like reefs assembled between them indicates the
uninterrupted reef development, at least in the marginal areas, during the
sedimentation. Unit 3 (Fig. 2) exhibits practically no reef. This is the last
lobe system that occurred during the so-called Messinian Salinity Crisis Event.
In Unit 2 (Fig. 2), the configuration of reefs (smaller, isolated reefs as well
as the scalloped arcs), reflects the configuration of the frontal profile of
the fan-delta lobes. The dimensions of the reefs are similar to the perimeter
of the lobes crest, while the depressions between them have been accurately
fossilized by the reef-talus deposits.

3 REEFS ASSOCIATED WITH DELTAIC FACIES

We have found two different situations in which reefs are associated with
different kinds of deposits originating in the deltaic environment. Both cases
may be studied in the Eocene deposits of the Catalan Basin. One of the cases is
also present in the Miocene Fortuna Basin.

In each one of these situations the reefs have different shape and
dimensions. The two kinds of reef-delta associations consist of: 1) great
arc-like fringing reefs developed on delta-front units, and 2) patch-reefs
positioned on stream-mouth-bars.

3.1 Arc-like Reefs Developed on Delta-front Units

In the first case the arc-like reefs are located on great sandstone units
of 5 km length and 100 m height. From an aerial view, these units have the
configuration of a lobe with great dimensions and a semicircular profile.

Extension of these units oscillate between minimal surfaces of 4.81 km^2 and 39 km^2. In longitudinal sections these sand bodies show a lenticular shape, wedging gradually in a proximal direction as well as in a distal direction.

The deposits that make up these units are characterized by thickening and coarsening upward sequences of great proportions (up to 100 m). These sequences do not show channelized facies at the top. In these sequences three areas may be distinguished, according to the kind of stratification that they present: 1) upper zone, with plane-parallel upper and lower contacts and plane-horizontal or weakly dipping stratification, 2) middle zone, with convex-concave contacts and planar or convex-concave cross-bedding, and 3) a lower zone with plane-horizontal or weakly dipping contacts and stratification in the same manner.

Due to the lack of channels in these sedimentary bodies, it is logical to believe that the distribution of matter was not made by means of a channelized current. The great extension that covers each one of the strata of this facies, its good grain-sorting, and the fact that it is made up of particles of the smallest size (fine - medium sandstone) point out a sedimentation mechanism formed by an unconfined current containing suspended matter. These currents must have been dispersed and diluted at the same time that the load was deposited. The front of the units constituted of this kind of facies is characterized by its strong depositional slope. The upper part of the units in the most proximal area formed a horizontal plane. Once the sedimentary activity ceased this area was converted into an ample platform on which shallow character sediments developed. On this platform, patch-reefs, stream-mouth-bars, and bars originated by tidal reworking developed. The existence and thick units described above are usually related proximally with a digitated stream-mouth-bar. Because of this and by their own characteristics, it might be inferred that these units are big delta-front units.

3.2 Relationships of Delta-front Units and Reefs

Along the margins of the Eocene Catalan Basin, there are great deltaic systems, whose volumetrically more significant section consists of a series of great coalescent delta-front lobes (Fig. 1B). On the frontal slope angle breaking point of these deltaic units, there is a fringing reef 40 km long and 800 m wide (Figs. 1B, and 6). From an aerial view this barrier reef has an arc-like pattern due to its adaptation to the delta-front lobes. There is a large platform of 10 km that has numerous patch-reefs. The reef-talus deposits are arranged in wedges that have adapted themselves to the delta-front slope (Fig. 6). These deposits are crossed by channels (filled by fragments of

skeletal origin), and these channels initiate in the fore-reef. Dimensions of
the reef-talus deposits depend mainly on the extension and gradient of the
delta-front.

The deltaic sedimentation in the Eocene Catalan Basin was developed in a
prograding direction in a centripetal manner from all the margins of the basin.
Each progradation stage was characterized by the formation of one or various
delta-front units (Fig. 6B). Fringing reefs were developed on those units. The
fossil record in this basin shows a great thickness of alternating deltaic
deposits and reefs in an excellent stage of preservation (Fig. 6A). Keeping in
mind this alternation in the context of the deltaic sedimentation, it may be
inferred that the development of the delta-front units was not simultaneous
with the reef growth at the same point. It was necessary for the delta-front to
become stabilized. From this we might point out that the reefs constitute the
abandonment facies of the delta-front units. However, the delta had
periodically renewed its activity (after the development of an arc-like reef)
following always the same pattern of matter distribution and forming its
deposits arranged in the same consistent manner. It indicates that the deltaic
system did not definitely lose its sedimentary activity. In fact, there was an
active deltaic sedimentation interrelated with the coralline growth of numerous
patch-reefs in the lagoon area during the development of the arc-like reefs on
the delta-front. In this way we have the record of the kind of interrelation
between deltaic deposits and reefs.

3.3 Reefs Developed in Relation with Stream-mouth bar Deposits

The second kind of deltaic deposits that are closely related with reef
buildups are composed by sandstones and conglomerates organized in thickening
and coarsening upward sequences. These sequences are usually cut at the top by
a conglomerate or a sandstone channel. The thickness of the sequence oscillates
between 3 m and 30 m.

The current structures that present these deposits vary also in a gradual
manner from the bottom to the top. At the lower part, thin (less than 5 cm)
sandstone layers prevail that alternate with mudstones and demonstrate a single
interval of isolated current ripples. In the middle part of the sequence, the
climbing ripples and the amalgamated current ripples with flaser structure are
common. At the upper part of the sequence the layers have a thickness that
oscillates between 40 cm and 1 m. These layers usually show cross-lamination
having its origin in the migration of minor dunes.

These deposits have been found arranged in bodies that reach 5 km in
longitudinal section and that transversally have dimensions that oscillate

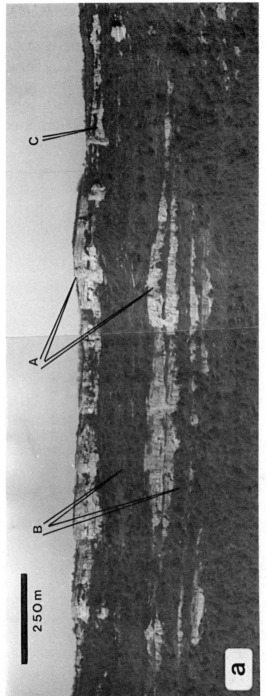

Fig. 6a. Reefs developed on deltaic deposits. Transversal section of various marginal arc-like reefs (A). (B) Delta-front units. (C) Talus-reef deposits.

48

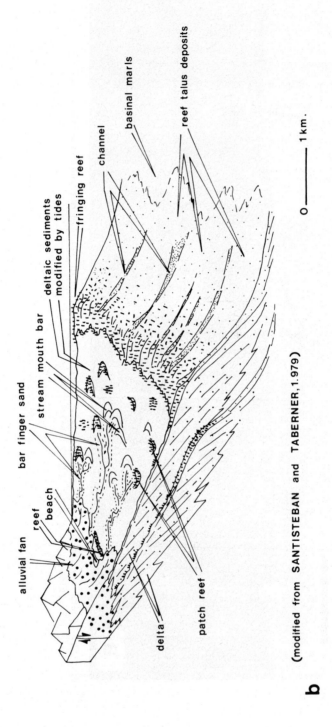

alluvial fan

reef

beach

bar finger sand

stream mouth bar

deltaic sediments
modified by tides

fringing reef

channel

basinal marls

reef talus deposits

delta

patch reef

0 ——— 1 km.

b

(modified from SANTISTEBAN and TABERNER, 1.979)

Fig. 6b. Reefs developed on deltaic deposits. Centelles Model.

between 800 m and 1 km. According to their dimensions these bodies have a tabular shape, elongated following the palaeocurrent direction.

This kind of deposit is located between deposits which originated in marine shallow water conditions, such as lagoon and interdistributary bay marls, and small coral reefs. They have been interpreted as deltaic units according to the structure that they present, and their relationships with other deposits. They began to deposit in the marine environment and at the front of the mouth of fluvial channels, and they continue developing by progradation of the channel (originating a digitate body). In relation to its shape and inner structure, these units are similar to the bar-finger sands described by Bates (1953), Fisk et al. (1954), Scruton (1956), Fisk (1961) and Gould (1970).

The reefs that are associated with the deltiac bar-finger sands adapted themselves to the external shape of these bodies. These reefs are positioned on the top and have adapted to the frontal slope of the bar (Fig. 7). The elongated shape of these reefs is partially conditioned by the deltaic sand bars. The outer reef deposits are arranged in wedge shaped forms whose angles attenuate at the frontal slope of the bar.

The reef-core is usually developed on a mixed carbonatic-siliciclastic level, which is rich in Nummalites and red algae. This kind of facies, Nummulites-rich facies, is also typical of the leeward side of these reefs. These kinds of deposits may be usually found resedimented and filling small channels 1.5 m in height and 25-30 m wide. These channels are located either between the talus-reef deposits or situated in depressions of the deltaic-bar frontal slope. The existence of these channels can be taken as proof in favor of the existence of a depositional slope in the front of the reef adapted to the deltaic bars. This deltaic depositional slope is suggested by the existence of a pronounced cross-bedding in the talus-reef deposits (Fig. 8). These kinds of reefs have varied dimensions. Their length (along the axis of the bar-finger-sand to which they have adapted) is between 100 and 800 m. The width is approximately 250 m. Due to its position and shape, from an aerial view it is possible to assert that these reefs are morphologically controlled by the external configuration of the deltaic bars.

From a strictly dynamic point of view, the development of these reefs cannot be considered contemporary with the advance of the deltaic lobes. However, these reefs do not represent the definite abandonment of the bar because, in general, the deltaic channel reestablishes its activity and the deposits of the new deltaic bar cover the reefs totally or partially (Figs. 7 and 8). This situation may be explained by the avulsion of the deltaic channels

50

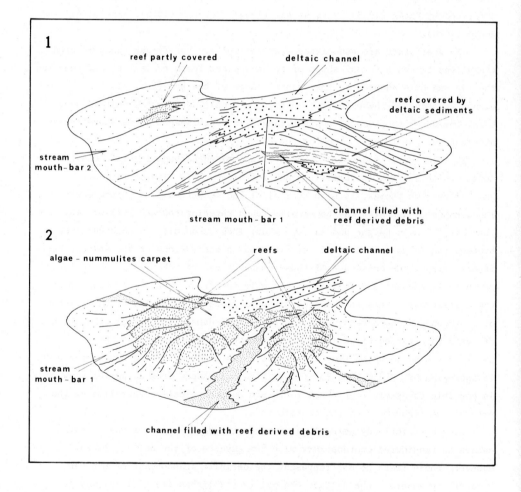

Fig. 7. Hypothetic reconstruction of the reefs associated with stream-mouth-bars. 1) The reefs colonize a stable stream-mouth-bar. 2) The reef itself has been covered by the deltaic bar deposits.

that have a tendency to migrate laterally, while part of the sediments already deposited are colonized rapidly by a reef community.

In the example that we have studied, the existence of deposits of alternating stream-mouth-bars-reefs is common. Stratigraphically both kinds of deposits seem to be interfingered; this is due to the existence of a continuous

Fig. 8. Patch-reef covered by deposits of a stream-mouth bar (S). The cross-stratification of the reef deposits corresponds to the depositional slopes of the reef-talus deposits (R).

reef growth on a deltaic platform where stream-mouth-bars are developed. This continuity has to be understood in the sense of time but not in that of space, since the migration of the deltaic channels and bars brought about the inundation of the reefs by the siliciclastics (Fig. 7). The relative elevation of the sea level allows the development of reefs on stream-mouth-bars related with a channel temporarily abandoned.

4 REEFS ASSOCIATED WITH CHANNELIZED TERRIGENOUS FACIES

In both studied basins, reefs present are also associated with channelized deposits. In this example presented, the dynamic relationships between the coralline growth and the terrigenous sedimentation is most closely linked.

Reefs present were developed on channels of deltaic platforms submerged due to a marine transgression. The reefs are located occupying the highest part of deltaic channels. Reefs are usually positioned in one of the margins of the channels and occupy half of the channel (Fig. 9). The other half of the channel is filled by siliciclastic sandstones. The reefs are part of the whole of channel infilling deposits, but they consist of in situ bioconstructions. In Figure 9, one may observe one of these reefs located at the margin of the channel. The reef-talus deposits are volumetrically more important than the area constituted by in situ corals. These deposits present inner planar cross-bedding that indicates the progradation direction of the reef towards the center of the channel. In each of these strata, there is a gradual augmentation of the percentage of terrigenous compounds that are present in the carbonate matrix. The other margin, not visible in Figure 9, is composed of sandstone layers deposited during the infilling of the channel.

This is the example that best reflects the existence of a terrigenous sedimentation developed during the coralline growth. The reef is positioned on a conglomerate channel lag deposit. The reef is covered by the sandstones of the final infilling of the channel (Fig. 9). In a general sense, the reef is stratigraphically equivalent to the sandstone deposits on the other margin of the channel.

It is interesting to point out that the field evidence suggests that the coralline growth continued during the periods in which there was sedimentation in the channel. However, it is possible that the cross-stratification surfaces of the reef, which represent stages of minor coralline development, might be equivalent to each one of the channel infilling sandstone beds. In this sense the large proportion of terrigenous compounds located at the top of each reef-talus layer is significant.

Through the study of this example of reef development inside a channel we can emphasize the following: 1) the channel was active during the reef growth, and 2) the coralline development was not completely interrupted during the periods in which there was transport and deposit of terrigenous sediments in the channel. It suggests the existence of a dynamic interaction between currents with load of terrigenous sediments and the reef itself.

5 CORAL REEFS RELATED WITH TIDAL BARS

An example of coral reef-terrigenous sand body association has been found in tidal deposits developed frontally to some stream-mouth-bars. These deposits have their origin in the reworking in the marine environment of the sediment previously arranged in stream-mouth bars. These deposits are composed of the superimposition of dunes of 10 m length and 40 cm height, which pass frontally into climbing ripples. Locally, associated with these deposits, there is hummocky cross-strafication.

Fig. 9A. (A) Axis of the channel. (C) Mixed clastic carbonate sediments. (D) Deltaic channel. (T) Reef-talus deposits.
(M) Margin of the deltaic channel. (R) Reef core.

54

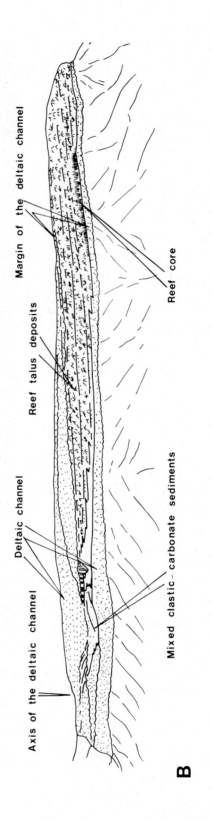

Fig. 9B. (A) Axis of the channel. (C) Mixed clastic carbonate sediments. (D) Deltaic channel. (T) Reef-talus deposits. (M) Margin of the deltaic channel. (R) Reef core.

These dunes are grouped in "greater" sand bodies (25-50 m length and 4m height). These "greater" sand bodies are separated from each other by small channels, whose axes are parallel to the paleocurrent direction shown in the dunes' grouping pattern. These deposits have shallow marine characteristics. At their base are lagoon marls (from the arc-like fringing reef). The top is usually colonized by coral reefs. In the proximal direction they pass gradually into stream-mouth-bar deltaic deposits. We have interpreted these deposits as a longitudinal bar system originated by tidal action. The disposition and geometry of these bars and channels are quite similar to those described by Johnson (1977) in the Late Precambrian of north Norway.

Coral reefs colonize the top of these tidal units. The reefs measure 3 m in height and generally develop on a group of tidal bars. The reef-talus deposits are arranged adapting to and alternating with the frontal slope of these bars. The coral reefs-tidal sand bodies association that we have studied is essentially morphological. We have not found evidence of a dynamic interaction between the two, but it is clear that the reefs have selectively taken advantage of the tidal sand bodies for their installation, in relation to the mudstones that surround the sand bodies.

6 REEFS ASSOCIATED WITH BEACH DEPOSITS

This kind of coastal reef is very well exposed in the Upper Lutetian-Biarritzian sediments in the Eocene Catalan Basin. We have studied several onlapping reef-beach stages in a transgressive situation.

The beach deposits are arranged in an ensemble of lineal outcrops located in the marine-continental transition of the southern and southeastern margin of the Vic Basin (belonging to the Eocene Catalan Basin) (Fig. 1B). The outcrops have a total approximate length of 100 km. On a large scale the beach deposits constitute tabular bodies. Their length, along the ancient shoreline, reaches 60 km. The greatest thickness of these beach deposits is shown in the central part of the outcrops, where there are eleven beach levels in a transgressive position, with a total thickness of 27 m. The lateral width of each beach unit is difficult to calculate because they are amalgamated with other beach levels. The maximum distance between the most proximal and the most distal areas of a beach is of 500 m. Thickness is usually not more than 4 m (Fig. 10).

The whole of the beach complex is located in the transition between alluvial continental deposits and marine sediments. In a distal direction beaches pass gradually into basinal marls. Most of the studied beach deposits correspond to constructive depositional beaches. They constitute tabular bodies with a longitudinal (proximal-distal) profile in a wedge that progressively attenuates towards the most distal areas. Their lower contact is usually planar and horizontal. The upper contact usually presents large scale undulations that correspond to the external shape of the small bar that constitutes the fore-shore. On a large scale the beach units are characterized by showing

Fig. 10. Reefs associated with beach deposits. (B) beach deposits. (D) shore-channel filled with mixed carbonate siliciclastic debris.

unidirectional planar large scale cross-bedding. The dip angle of the inner stratification surfaces oscillates between a minimum of 5° and a maximum of 15°. The direction of inclination of these surfaces is toward the open sea.

There are coral reefs associated with beach deposits (Figs. 10 and 11). Reefs may be located in three different stratigraphic situations in relation with the beach deposits: 1) reefs placed at the front of beaches, 2) reefs developed on beaches, and 3) reefs covered by beach deposits.

At first reefs are developed in a frontal position in relation with the beach, and after that, they cover the fore-shore deposits (Fig. 10). It may be considered as an effect of the Lutetian-Biarritzian transgression. It has also manifested itself through the step-like displacement of the ensemble of reef and beach levels towards land (Fig. 11). The transgressive migration of reefs occurred simultaneously with that of beaches (Fig. 11). In cases where the beaches are presented as isolated units, the reefs are also isolated in a like manner. When the beaches in a transgressive position form a single level, reefs are present in a continuous carbonate layer covering the beach deposits (Fig. 10).

The parallel evolution of reefs and beaches suggests that: 1) reefs grew in front of the beaches and both developed simultaneously, 2) reefs were dynamically related with the beach terrigenous sediments, and 3) reefs

57

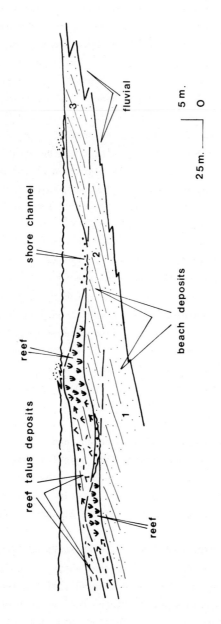

Fig. 11. Sketch of facies associations with beach deposits. Beaches in transgressive position.

positioned themselves on ancient submerged levels (Fig. 11). In various
localities where reefs have been found developed frontally to the beaches one
may distingish (from sea to shore): 1) reef buildup, 2) morphological
depression between the reef and the beach, and 3) the terrigenous beach. The
reef growth usually began at first on the fore-shore break in the slope of an
ancient submerged beach and after that extended backwards as a response to the
marine transgression (Fig. 11). The reef-talus deposits originated covering the
fore-shore deposits of the ancient submerged beach. The morphological
depression may have been preserved in the fossil record as a depressed area as
such or as a sandstone conglomerate infilling channel. In some cases the
channel infilling deposits present seawards inner cross-bedding (Fig. 10). The
cross-bedding must have been originated by the migration of the morphological
depression because of the beach progradation. The depression infilling consists
of a mixed facies, whose compounds are fragments coming from its own reef, and
also from terrigenous elements.

The reef-beach association is repeated numerous times in the deposits
which register the Lutetian-Biarritzian marine transgression. There are two
kinds of relationships that may be deduced from this terrigenous-carbonate
association: 1) morphologic, and 2) dynamic. The morphologic relation consists
of the control exerted by the external shape of an already submerged beach on
the installation point of the reef community. The dynamic relation is deduced
from the existence of a reef community in front of an active beach.

The littoral sedimentation at the shore, and sand transport along the
shore channel and the turbulent removal of the sediment by the waves in the
inner areas of the reef are the most effective controls for the inshore
development of the coastal fringing reef.

In the area where reefs are related with beach deposits the best preserved
beach deposits of the basin have been found. This suggests that, at least
partially, these reefs could have protected the beach from its destruction by
marine action. This protection could have been of two principal types: physical
protection from the waves, and stabilization of the already drowned beaches.

7 SILICICLASTIC TURBIDITES CONDITIONED BY THE OUTER SLOPE OF THE REEF-TALUS

In most above mentioned examples the external shape of the sedimentary
body passively controlled the location and arrangement of the reef community.
There are however, other cases in which a fringed-reef buildup developing on a
terrigenous sedimentary system can modify the flow conditions of the current
loaded with matter. Therefore the reef may condition the formation of the
terrigenous deposits. We have studied one of the these examples in the Eocene
Catalan Basin. It consists of the formation of turbidite deposits due to the

shift along the fore-reef slope of matter that originates from a deltaic
system. In this example the terrigenous sedimentation developed during the reef
growth. In spite of this the reef does not present signs of the negative effect
of the currents carrying a load of terrigenous sediments. The reef does not
show inner erosion surfaces nor levels in which the corals present evidence of
having been limited by the siliciclastics during their growth. The result of
the contemporaneousness of the terrigenous sedimentation and the coralline
development consists in this case, of a mixed deposit constituted by reef
carbonates that interfinger towards the lagoon area and also towards the
fore-reef, with siliciclastic sediments. This model is based primarily on the
existence of a marginal arc-like reef similar to the one previously described
in "reefs associated with deltaic facies", that develops having adapted itself
to the front of a delta front unit (Fig. 6). In this case the lagoon deposits
are very reduced, probably due to the existence of a rather unstable deltaic
platform and a more important deltaic sedimentation.

The reef is arranged in prograding position (Fig. 12). The ascending
arrangement of the reef core and the other facies of this reef suggest a
continuous relative ascent of sea level.

In the reef core there are neither terrigenous deposits nor mixed facies,
in spite of the infilling of some channels located in the reef wall. In the
outer reef area there are three main accumulations of sediments which
originated in the reef in a wedge-in-wedge deposition (Fig. 12). The three
units are composed of coral rubble and calcarenitic deposits. The wedges
correspond to the three more effective progradation stages of the reef. The
talus-reef deposits present are interfingered with siliciclastic deposits (Fig.
12). The siliciclastic sediments constitute wedge shape units that attenuate
towards the reef. These units of terrigenous matter are arranged through a
non-erosive net contact of the talus-reef deposits. Only in one case are they
directly deposited on the reef core front. Once the outcrops in the field have
been studied, one may conclude that there is no physical connection between the
siliciclastic deposits interfingered with reef-talus deposits and terrigenous
deposits present in the lagoon area. The siliciclastic units show turbidite
facies associations similar to those described by Mutti and Ricci-Lucchi (1972)
and Mutti (1977). Each one of the units presents basically the same facies
associations distributed in the same consistent manner. These facies
associations may be genetically grouped in two different ensembles: 1)
turbidites in continuous beds, and 2) associated dunes and ripples.

The turbidites in continuous beds have a thickness which oscillate between
1 and 20 cm. These layers may be arranged isolated between mudstones or in
groups of beds measuring 1.5 m. The mudstone/sandstone ratio is variable but

60

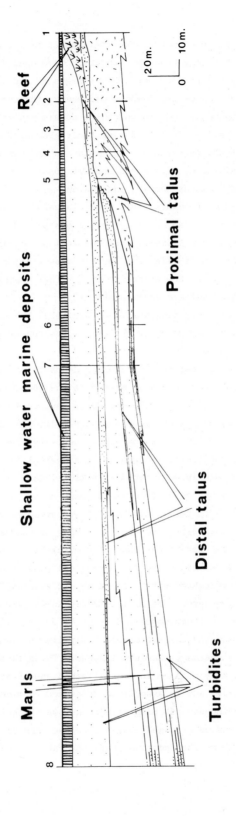

Fig. 12. Stratigraphic cross-section of turbidite deposits associated to reefs.

usually is high (10/1). The turbidite beds present B, C, and D intervals of the Bouma sequence (Bouma, 1962). Generally in the inner sandstone levels the interval C and D dominate with respect to the B, which may be absent. In general, in the thicker sandstone levels the interval C is very developed. This interval C is constituted by climbing ripples.

This facies association is characterized by the existence of slump deposits and slump scars. The slump scar structures are more abundant than the slump deposits, so that we could conclude that the association of turbidites in continuous beds was developed in an area where the slumping processes were originated. Slump scars are always fossilized by the facies association. In this facies association there are no channelized deposits. Only some layers present erosive bases. Because of the absence of channels we do not have evidence that the turbidity currents which originated these layers were channelized.

The facies association formed by associated dunes and ripples is characterized by sand bodies enclosed in mudstones. These units of sandstones have variable dimensions (from 2 to 15 m length, and 30 cm to 2 m height) and an external bar shape. Each one of these bodies is composed of some minor units consisting of dunes corresponding with the facies association B_2-E from Mutti and Ricci-Lucchi (1972). These minor units have lengths less than 10 m and heights of 40 cm. In each of them the dunes represent the more distal area, in the direction of the palaeocurrent. The continuous ripples and lenser occupy the proximal area. The dunes on occasions have an internal anti-dune structure.

The sand bodies of this facies association consist of various minor units in a cross-current off-lap position. These bodies were developed by vertical accretion. The result of this positioning is the formation of bars with thickening and coarsening upward sequences. The sedimentary current structures that this bar sequence present are, from bottom to top: isolated ripples, continuous ripples, amalgamated ripples with a flasser structure, dune cross lamination and anti-dune lamination. It is common for these associated dune units and ripples to be located on top of each other in a cross-current overlapping position. The contacts between each one of these units are erosive. These units must have been formed through a mechanism that combined erosion and sedimentation and that migrated in an upward current. The structure that this type of positioning presents suggests that the deposits form under upper flow regimes, through the migration of chutes and pools (Simons et al., 1965; Hand, 1974).

The two types of facies associations described present themselves in every case as being positioned directly on the talus deposits. Each one of these associations occupies a fixed place along the length of the slope of the reef-talus (Fig. 13).

The association of turbidites in continuous layers with slump scars always is presented on the upper part of the talus (proximal talus). The association formed by dunes and ripples occurs always on the distal talus, or in the transition between the proximal talus and the distal talus (Fig. 13).

Along the slope of the talus wedges of the reefs we have studied three zones with distinct gradients can be distinguished. The base of the system of buttresses in the uppermost part of the upper talus show a gradient of 0.19-0.3. The upper talus has a gradient of 0.07. The corresponding value of the lower talus varies between 0.008 and 0.01.

The fact that each one of the facies associations described is always presented on the same area of the talus-reef suggests the existence of a genetic relation between them and the gradient angle of the substratum. The facies association formed by turbidites in continuous layers must have been originated by unconfined turbidite currents starting from the foot of the buttresses (Fig. 13). This facies association does not contain channels. The existence of a great quantity of siliciclastic matter in the channels of the spurs and groove system of the reef suggests that these materials were introduced in the basin through the drainage channels of the reef. The turbidite currents that deposited this facies association could have been developed under natural conditions, in a similar way in which artificial currents were developed in a laboratory by Luthi (1981). In both cases, the deposits formed present the same sedimentary structures. In Luthi's experiment a depositional surface with a constant gradient of 0.085 was used. This value is similar to a proximal reef-talus of those studied by us (0.07). In our case, the angle of inclination was sufficient for numerous slumpings to be produced.

The facies association B_2-E is presented in the field apparently disconnected from the deposits formed by continuous turbidite layers. Between both facies associations there exists a pelitic zone lacking sandstone layers (Fig. 13).

The fact that bars formed by these facies associations are not related to a channel system suggests that they were originated by a non-channelized current developed along the superior talus-reef slope. These bars are found generally in the zone where the angle changes between the upper talus and the lower talus (Fig. 13). These two facts (the absence of channel and the presence of a pronounced decrease in the angle of the substratum) suggests the existence of a by-passing sedimentary zone in which a reworking of the bottom and bar formations of a facies association B_2-E was produced (Mutti, 1974). Bars having a similar structure and which are composed of the same facies association have

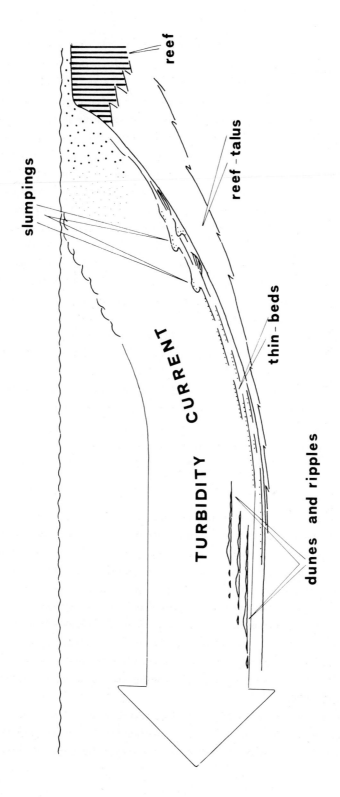

Fig. 13. Idealized model of the formation of turbidity currents by the shifting of deltaic matter along the slope of a reef-talus.

been previously described by Mutti (1974, 1977) in the Hucho Group in the
southern Eocene Pyrenean Basin. This author interprets these deposits as
suprafan lobes or channel mouth bars.

The two facies associations described from the deposits of non-fan shaped
turbidite sediments are presented forming wedge shape units which are
interfingered between reef-talus deposits (Fig. 12). Each of these units
correspond to a unique turbidite system. In this example, there are various
types of depositional relations existing between the reef development and the
turbidite formation. 1) The reef buildup conditioned the formation of turbidity
currents. The deltaic sediments deposited in the lagoon area are drained
outside of the coralline nucleus through the channel system of the reef front.
The turbidite currents originated at the mouth of these channels probably due
to the steep angle of the reef front (Fig. 13). 2) The most important changes
in the flow regime of these turbidite currents are conditioned by the changes
in the angle of the upper surface of the talus-reef. 3) The turbidite
sedimentation is contemporary with the reef growth. The corals in the reef
continued developing during the turbidite formation. These corals survived the
deltaic sedimentation in the lagoon area and the passage of this sediment
through the reef core.

In this example one may also differentiate two types of relationships
between siliciclastic deposits and reefs: 1) passive control by the external
form of a reef over the formation and positioning of the siliciclastic facies;
2) a dynamic relations, the result of the contemporaneousness of the
terrigenous sedimention and the coral growth.

8 CONSIDERATIONS

The existence of siliciclastic sedimentation is considered incompatible
with reef growth. Nevertheless, in the fossil record numerous examples exist of
fossil reefs that developed in active systems of terrigenous sedimentation. In
this paper we have described some of these examples in which there is evidence
of contemporariness or of great proximity in time between coral reef growth and
siliciclastic sedimentation. Also, some cases of living reefs situated in
active fan-delta systems are known (Gvirtzman and Buchbinder, 1978; Guilcher,
1979; Hayward, 1982; - in the Gulf of Aqaba - ; and Wescott and Ethridge, 1980,
in the Yallahs fan-delta, Jamaica), as well as in tidal channels (Evans et al.,
1973; Loreau and Purser, 1973), or fringing the front of a tidal delta (Loreau
and Purser, 1973).

One of the most principal problems that presents itself when we consider
the relationships between terrigenous sedimentation systems and the reefs is
the complexity of the exact nature of these relationships. We have

distinguished two types of relationships by reason of the positioning of the siliciclastic deposits and reefs in the fossil examples studied: 1) dynamic relationship, 2) morphological relationship. The existence of a dynamic relationship implies a proximity in time and space between reef development and terrigenous sedimentation. The simultaneousness between coral growth and terrigenous sedimentation implies a maximum dynamic relation. There is no dynamic relation when the siliciclastic deposits and reefs are presented completely separated in time and space.

We have found a sedimentary model that unites various examples of relationships between terrigenous carbonates previously described. This model is based on the outcrops of the Eocene deposits found around the village of Centelles in the Catalan Eocene Basin (Fig. 1B). Because of this location, the model has been named the Centelles model (Fig. 6B) (Santisteban and Taberner, 1979, 1980). On this Centelles model, reefs are related to fan-delta lobes, beaches, tidal bars, deltaic channels, stream-mouth-bars, and delta-front lobes. If we consider only the relationships of the reefs with the deltaic deposits we are able to recognize that three types of dynamic relationships exist.

The Centelles model (Fig. 6B) consists "a grosso modo" of a prograding system of large superimposed delta front units, on which large marginal arc-like reefs developed (Fig. 6). Each arc-like reef usually is positioned on one or various coalescent delta front units. These large arc-like reefs coincide with the formation of the delta front units, in space but not in time, since they are situated on stabilized deltas. Only in the example describing turbidites conditioned by the slope of the talus-reef is there a record of an active deltaic siliciclastic sedimentation that reaches the far edge of the deltaic units when the reef is originated. The marginal arc-like reefs are separated by the length of time that a new delta front unit need to develop. There existed in the lagoon area active siliciclastic sedimentation in the form of fan-delta lobes, stream-mouth-bars, beaches, and tidal bars. Patch reefs developed attached to these sedimentary bodies. Two types of relationships between these patch reefs and the fan-delta lobes or stream-mouth bars have been found. 1) Both coincide in time but not in space. 2) Both coincide in space but only in short periods of time. The interval of interruption of reef growth is much less than that which exists between two arc-like reefs. Only in one case have we found in this model a deltaic channel in which there exists evidence of contemporariness between reef growth and terrigenous sedimentation. In this case, just as in the case of the turbidites conditioned by the slope of a talus-reef, there exists coincidence in time and space.

The morphological relationship consists of the control by the external form of sand bodies over the reef development or vice-versa. In the majority of the cited examples the reefs are adapted to bar or lobe shaped bodies. The most notable feature of this adaptation is the colonization of inclined substrata by coral.

The greatest part of the fossil reefs that we have studied corresponds to communities with high species diversity, with a very well defined species zonation, and with a very clearly defined zonation of morphologies of corals that show polymorphism. Some functional modifications of the coral caused by the incidence of determined physical factors on the reef community are imprinted in the in situ structure of the coral that show polymorphism. In ecologically mature reefs not having an excessive pressure by determined conditioning factors (currents, siliciclastic matter, etc.) the coral is better adapted to greater effectiveness in capturing daylight. The coral adapt to capture daylight mainly by means of modification of their surfaces and as a result, their entire form is modified. Other modifications are due to the effect of currents over some parts of the reef (Geister, 1977). Yet others are the result of the competition for acquiring nutrients.

The functional adaptations of the coral facilitates the creation of smaller and more specific compartments in the reef. It also conditions important changes in the configuration of the environments (current deflection and attenuation, establishment of micro-environments). Figure 14 illustrates the fields of variability of different physical factors that act on coral reefs that grow on a previously existent morphology. There exists a large quantity of factors that can influence coral growth, but here we have indicated only some of them. The factors that vary vertically produce a horizontal stratification of the coral morphologies. The other factors that vary horizontally cause the positioning of the coral morphologies in bands parallel to the coast. The sum of these factors causes different results in each point of the reef. Its effectiveness is determined in some way by the angle of the substrate. In the Miocene reef of the Fortuna Basin, which is adapted to fan-delta lobes, Porites lobatosepta and Tarbellastraea eggenburgenis present a great variety of coral morphologies (Figs. 15 and 16). Both corals show the same morphological varieties in the same reef position. In the reefs studied any of these two species can be dominant. In these reefs there exists an increase throughout time in diversity of the species. If, for example, the dominant coral is Porites this coral is the first to colonize in head shape and dish-like colonies (Figs. 15 and 16). The zone in which the coral develop first is the breaking point of the slope (Figs. 15 and 17). Later the colonies extend towards the crest of the bar and the most shallow areas. In this stage

67

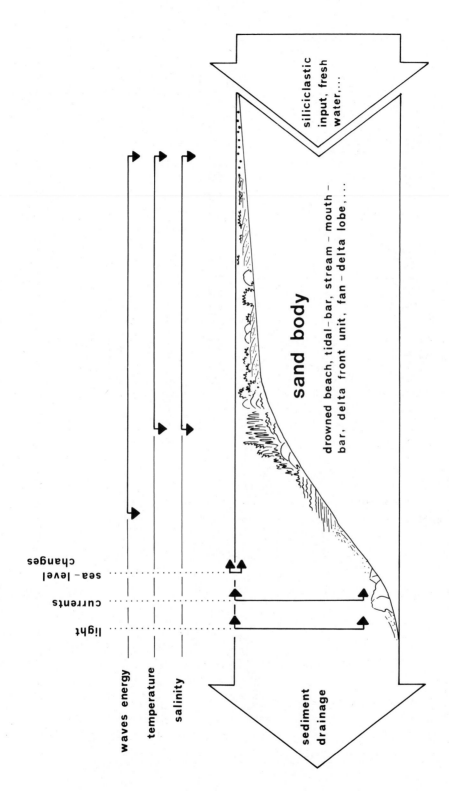

Fig. 14. Sketch that shows the fields of effective action of various physical factors on a reef community adapted to the front of a lobe or bar shaped body.

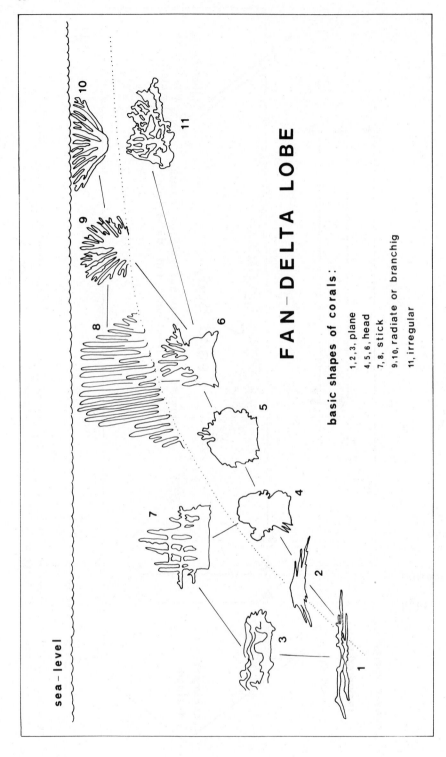

Fig. 15. Basic morphologies of *Porites* and *Tarbellastraea* and their position in a Miocene reef adapted to the front of a fan-delta lobe.

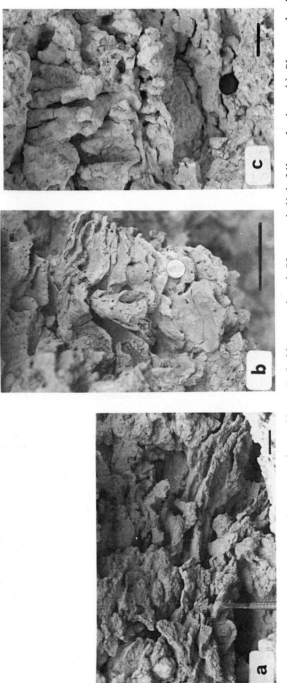

Fig. 16. Basic morphologies of Porites (as well as Tarbellastraea): a) Plane and dish-like colonies. b) Plane colonies with digitated protuberances. c) Bell shaped colonies with digitated protuberances. See their position along the frontal slope of the fan-delta lobe in Figure 15. Scale bars are 10 cm in all photographs.

70

Fig. 16. Basic morphologies of _Porites_ (as well as Tarbellastraea): d) Sticks. e) Head shaped forms. f) Head shaped forms with digitated protuberances. See their position along the frontal slope of the fan-delta lobe in Figure 15. Scale bars are 30 cm in "d" and 10 cm in photographs "e" and "f".

Fig. 16. Basic morphologies of <u>Porites</u> (as well as <u>Tarbellastraea</u>): g) Radiating colonies. h) Irregular shapes. See their position along the frontal slope of the fan-delta lobe in Figure 15. Scale bars are 10 cm.

72

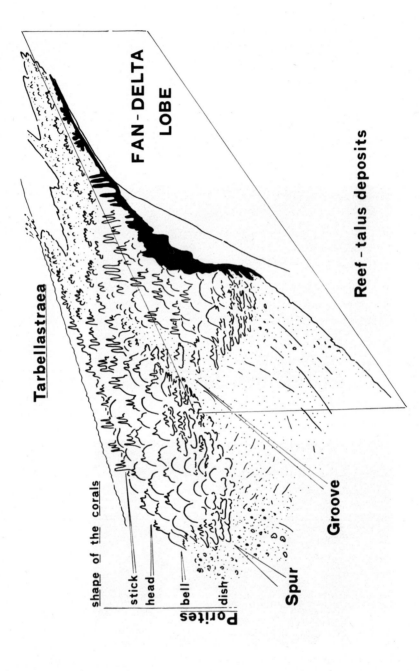

Fig. 17. Reef configuration and zonation of coral morphologies of a Miocene reef adapted to the front of a fan-delta lobe.

Montastraea sp. and Diploria sp. appear in the upper talus. Lastly
Tarbellastraea develops in the uppermost parts of the reef with branching
colonies (Fig. 17).

The density of coral skeletons is higher in this type of reef developed on
inclined surfaces than on horizontal planes. According to Loya (1972) the reefs
of the Gulf of Elat, adapted on the front of a fan-delta, also present a great
species diversity and a high percentage of living corals.

We suggest that it is possible for substrata with elevated gradients to
facilitate small disturbances caused by instability. According to the ideas of
Connell (1978) these disturbances can serve as a stimulus for maintenance of
high species diversity in these reefs. Based on these examples studied, we
believe it possible to suggest the hypothesis that the non-catastrophic
disturbances produced by the sedimentary activity in a fan-delta lobe, channel,
etc., can be assimilated by the reef and result positively for the maintenance
of a high coralline diversity. The existence of spasmodic currents loaded with
the sediment which is deposited on the reefs has the effect of preventing the
elimination of inferior competitors (Connell, 1978). As a consequence the reef
community adapts more to the physical external agents and takes a long time to
reach a climax state.

The fact that the coral community is developed on a steep slope may
explain the attenuation of the effect of the sediment deposited among the
corals. This is explained due to the fact that a steep slope can favor the
drainage of sediment. These reefs adapted to fan-delta lobes seem to achieve
the greatest effectiveness in drainage. These reefs consist almost exclusively
of a system of channels and buttresses. The effectiveness of drainage is
manifested in the great quantity of terrigeneous sediment deposited in the
reef-talus, while it is very scarce in the matrix of the buildup.

9 CONCLUSIONS

In the fossil record we frequently find interfingered reef buildups with
siliciclastic deposits. In some cases these reefs developed in active
terrigenous sedimentary systems. One may distinguish two types of relationships
between reefs and sedimentary bodies: 1) a dynamic relationship which responds
to proximity in time and space between reef growth and sedimentary activity; 2)
a morhpological relationship. The morphological relationship consists
principally of the conditioning of the reef by the form of the substrata.

The sedimentation of terrigenous materials in a reef (when there are no
destructive factors) is a physical factor that can control the external
configuration in the reef community.

74

10 ACKNOWLEDGEMENTS

We are grateful to Dr. R.N. Ginsburg and Dr. L.J. Doyle for having suggested to us the possibility of presenting this paper and for having been of invaluable help and support.

The authors wish to acknowledge Professor A. San Miguel from the Department of Petrology of the University of Barcelona for making available the facilities at the department.

We also would like to thank our colleagues at the Department of Geology (University of Valencia), especially Professor M. De Renzi and Dr. A. Marquez-Aliaga, whose aid has been immeasurable.

Diane Bax has performed the laborious task of translating this paper into English.

This paper has been fulfilled during a period of time established by a grant (Reintergro), through the Spanish Ministry of Education and Science.

11 REFERENCES

Bates, C.C., 1953. Rational Theory of Delta Formation. Am. Assoc. Petroleum Geologist Bull. 37(9): 2119-2162.

Bouma, A.M., 1962. Sedimentology of some flysch deposits. A graphic approach to facies interpretation. Amsterdam, Elsevier Publ. Co., 168 pp.

Brown, L.J., Jr., 1969. Geometry and distribution of fluvial and deltaic sandstones (Pennsylvanian and Permian), North Central Texas. Gulf Coast Assoc. Geol. Socs. Trans., 19: 23-47. Reprinted as Texas Univ. Bur. Econ. Geology, Geol. Circ., 69 - 4.

Brown, L.F., Jr. and Fisher, W.L., 1977. Seismic Stratigraphic Interpretation of Depositional Systems. Examples from Barzilian Rift and Pull-Apart Basins, In: Ch. E. Payton (Editor), Seismic Stratigraphy Applications to Hydrocarbon Exploration. Am. Assoc. Petroleum Geologists, Mem. 26: 213-248.

Brown, L.F., Jr., Cleaves, A.W., II and Erxleben, A.W., 1973. Pennsylvanian depositional systems in North Central Texas, a guide for interpreting terrigenous clastic facies in a cratonic basin. Texas Univ. Bur. Econ. Geology Guidebook, 14: 132.

Choi, D.R. and Ginsburg, R.N., 1982. Siliciclastic foundations of Quaternary reefs in the Southernmost Belize Lagoon, British Honduras. Geol. Soc. of America Bull., 93: 116-126.

Choi, D.R. and Holems, Ch. W.E., 1982. Foundations of Quaternary reefs in south central Belize Lagoon, Central America. Am. Assoc. Petroleum Geologists Bull. 66(12): 2663-2681.

Connell, J.H., 1978. Diversity in tropical rain forests and coral reefs. Science, 199: 1302-1310.

Dabrio, C.J., 1975. La sedimentacion arrecifal neogena en la region del rio Almanzora. Estudios Geologicos, 31: 285-296.

Evans, G., Murray, J.W., Biggs, H.E.J., Bate, R. and Bush, P.R., 1973. The oceanography, ecology, sedimentology, and geomorphology of parts of the Trucial Coast Barrier Island Complex, Persian Gulf. In: B.H. Purser (Editor), The Persian Gulf. Berlin, Springer-Verlag, pp. 279-328.

Fisk, H.N., 1961. Bar-finger sands of Mississippi delta. Geometry of Sandstone Bodies. Am. Assoc. Petroleum Geologists. Symposium Volume, pp. 29-52.

Fisk, H.N., McFarlan, E., Jr., Kolb, C.R. and Wilbert, L.J., Jr., 1954. Sedimentary framework of the modern Mississippi delta. Jour. Sed. Petrology, 24(2): 76-99.

Friedman, G.M., 1982. Coexisting terrigenous sea-marginal fans and reefs of the shore of the Gulf of Aqaba. Int. Assoc. of Sedimentologists. Eleventh International Congress on Sedimentology. McMaster University, Hamilton, Ontario, Canada. Abstracts of Paper, pp. 109.

Geister, J., 1977. The influence of wave exposure on the ecological zonation of Caribbean coral reefs. Proc. 3rd Int. Coral Reef Symp., Miami, 1: 23-29.

Gould, H.R., 1970. The Mississippi delta complex. In: J.P. Morgan and R.H. Shaver (editors), Deltaic Sedimentation, Modern and Ancient. Soc. Econ. Palaeontologists Mineralogists Spec. Publication, 15: 3-30.

Guilcher, A., 1979. Les rivages coralliens de l'est et du sud de la presqu'ile du Sinai. Ann. Geographic, 488(58): 394-418.

Gvirtzman, G and Buchbinder, B., 1978. Recent and Pleistocene Coral Reefs and Coastal Sediments of the Gulf of Elat. Tenth International Congress on Sedimentology. Jerusalem Post-Congress Excursion Y, 4: 163-191.

Hand, R.M., 1974. Supercritical flow in density currents. Jour. Sed.Petrology, 44(3): 637-648.

Hayward, A.B., 1982. Coral Reefs in a Clastic Sedimentary Environment. Fossil (Miocene, S.W. Turkey) and Modern (Recent, Red Sea) Analogues. Coral Reefs, 1: 109-114.

Johnson, H.D., 1977. Shallow marine sand bar sequences: an example from the Late Precambrian of North Norway. Sedimentology, 24: 245-270.

Loreau, J.P. and Purser, B.H., 1973. Distribution and Ultrastructure of Holocene Ooids in the Persian Gulf, In: B.H. Purser (editor), The Persian Gulf. Berlin, Springer-Verlag, pp. 279-328.

Loya, J., 1972. Community structure and species diversity of hermatipic corals of Elat, Red Sea. Marine Biology, 29: 177-185.

Luthi, S., 1981. Experiments on non-channelized turbidity currents and their deposits. Marine Geology, 40(3/4): M59-M68.

Mutti, E., 1974. Turbiditas de Suprafan en el Eoceno de los alrededores de Ainsa (Huesca). VII Congreso del Grupo Espanol de Sedimentologia. Bellaterra - Tremp. Resumenes de las comunicaciones, pp. 68-71.

Mutti, E., 1977. Distinctive thin-bedded turbidite facies and related depositional environments in the Eocene Hecho group (South Central Pyrenees, Spain). Sedimentology, 24: 107-131.

Mutti, E. and Ricci-Lucchi, F., 1972. Le torbiditi dell'Appennino settentrionale. Introducion all'analisi di facies. Mem. Soc. Geologica Italiana, 11: 161-199.

Reid, P. and Tempelman-Kluit, D., 1982. An Association of Reefal Carbonates and Volcano-Clastics in the Upper Triassic of the Yukon Territory, Canada. Int. Assoc. of Sedimentologist. Eleventh International Congress on Sedimentology. McMaster University, Hamilton, Ontario, Canada. Abstracts and Papers, pp. 110.

Santisteban, C. and Taberner, C., 1977. Barras de marea como control de la formacion de arricifes en el Eoceno medio y superior en el sector de St. Feliu de Codines-Centelles. Inst. Inv. Geol. Diputacion Provincial. Universidad de Barcelona, 32: 203-214.

Santisteban, C. and Taberner, C., 1979. Relacion entre sedimentos terrigenos costeros, facies arricifales y evaporitas. El modelo de Centelles y su aplicacion regional. Act. Geologica Hispanica. Homenatge a Ll. Sole Sabaris, 14: 229-236.

Santesteban, C. and Taberner, C., 1980. The siliciclastic environments as a dynamic control in the establishment and evolution of reefs. Sedimentary models. Int. Assoc. of Sedimentologists. 1st European Regional Meetings. Bochum. Abstr., pp. 208-211.

Scruton, P.C., 1956. Oceanography of Mississippi delta sedimentary environments. Am. Assoc. Petroleum Geologists Bull., 40: 2864-2952.

Simons, D.B., Richardson, E.V. and Nordin, C.F., Jr., 1965. Sedimentary
    structures generated by flow in alluvial channels. Soc. Econ.
    Palaeontologists Mineralogists Spec. Publication, 12: 34-52. In:
    Sedimentary Processes. Hydraulic Interpretation of Primary Sedimentary
    Structure (compiled by G.V. Middleton): Soc. Econ. Palaeontologists
    Mineralogists Reprint Series No. 3 (1977).
Wescott, W.A. and Ethridge, F.G., 1980. Fan-delta Sedimentology and Tectonic
    Setting - Yallahs Fan-delta, Southeast Jamaica. Am. Assoc. Petroleum
    Geologists Bull., 64(3): 374-399.

Chapter 3

CASE HISTORIES OF COEXISTING REEFS AND TERRIGENOUS SEDIMENTS: THE GULF OF
ELAT (RED SEA), JAVA SEA, AND NEOGENE BASIN OF THE NEGEV, ISRAEL

G.M. FRIEDMAN
Department of Geology, Brooklyn College of the City University of New York
Brooklyn, New York 11210, and Rensselaer Center of Applied Geology
Affiliated with Brooklyn College of the City University of New York, 15
Third Street, P.O. Box 746, Troy, New York 12181

ABSTRACT
     The contemporaneous deposition of clastics and development of
carbonate-secreting organisms, once thought to have been mutually exclusive,
is actually quite common through a variety of sedimentological and
biological processes.
     In the Gulf of Elat (Red Sea), the hot, dry climate, and infrequent
rain-fall result in flash-flooding. Coarse sands, washed down from the
adjacent highlands through intermittant streams (wadis), are deposited in
alluvial-fan complexes while reef complexes contemporaneously are deposited
along the shelf break. The carbonate-secreting organisms overcome the
inhibiting effects of the clastic material due to sediment coarseness and
the long periods of quiesence between flash-floods. Reef complexes almost
fringe the entire Gulf of Elat coastline.
     The humid climate of the Java Sea, contrary to the Gulf of Elat, ensures
a consistently heavy, sediment-ladened run-off from the islands composed
primarily of fine-grained material. Longshore currents redistribute these
sediments laterally, away from the carbonate-secreting organisms allowing
reef development in clear water approximately 25 km from the Java coast.
     Within the Hazeva Formation in the Neogene Yeroham Basin in Southern
Israel, terrigenous and carbonate facies are mixed in an upward-coarsening
clastic sequence topped by an erosion-resistant oyster bank (Crassostreids).
By analogy with modern environments, the oyster bank is considered to be
intertidal in origin. Probably as a paleoecological retreat into a setting
where predators cannot survive, oysters have adapted to a muddy environment
by eliminating clay-mucous particles as pseudo-feces. Here, the mutually
exclusive relationship between fine-grained terrigenous sediments and
carbonate-secreting organisms is not the case.

1 INTRODUCTION

     Carbonate sediments form many of the world's major petroleum reservoirs.
To locate new carbonate petroleum reservoirs, modern carbonate depositional
environments have been studied in attempts to reconstruct ancient
depositional settings. This approach has led to the observation that
carbonate-secreting organisms and terrigenous sediments may coexist. The
prevailing theme in sedimentology has been that terrigenous deposits inhibit
carbonate production. Most carbonate-secreting organisms are filter feeders,
hence carbonate sedimentation has been considered clear-water sedimentation.
However, recent studies (references) show that coexisting terrigenous and
carbonate sediments are actually quite common.

The mixing of terrigenous sediments and carbonate material involves a variety of sedimentologic and biologic processes. Mount (1984) has categorized the mixing processes in four ways: (1) "punctuated mixing" or transferring of sediment between contrasting depositional environments during rare, high intensity sedimentation events, (2) sediment mixing along the contact between contrasting facies, (3) in situ mixing through the autochthonous generation of carbonate material within siliclastic sediments, and (4) through the erosion of uplifted carbonate source terrains. This paper will examine modern examples of coexisting terrigenous and carbonate sediments [Gulf of Elat (Red Sea), Java Sea reef tracts] as well as an ancient one (Yeroham Basin in southern Israel).

2 MODERN ANALOGUES

2.1 The Gulf of Elat

A constant influx of fine-grained terrigenous material will inhibit carbonate production, however major storm events may result in the periodic transfer of large volumes of sediment from one facies to another. The transferring of sediment during rare, high-intensity sedimentation events as well as in situ mixing and erosion of exposed carbonate-source terrains is vividly demonstrated in the Gulf of Elat (Friedman, 1968) (Fig. 1). Reef complexes are deposited along the shelf break while clastics are contemporaneously deposited in alluvial-fan complexes which spill out onto a narrow shelf and in deep-sea fan (turbidite) complexes.

The climatic conditions in the Gulf of Elat strongly affect sedimentation processes in and marginal to the Gulf with respect to the influx of terrigenous sediment, the accumulation of supratidal evaporites and algal-mat carbonates, and intrabasinal carbonate deposition associated with fringing coral reefs and pelagic organisms. Hot and dry with a mean annual air temperature of 25.7° and an average rainfall during a twenty-year period of approximately 25 mm (Red Sea and Gulf of Aden Pilot, 1967; Lebedev, 1970), most of the rainfall in the Gulf of Elat occurs within just a few days; during some years there is no rainfall at all. When rain does fall catchment areas of up to tens of square km in areal extent in the rocky desert highlands funnel flash floods through wadis (intermittent stream valleys) out to the narrow coastal margins of the Gulf. The fans build out as much as several kilometers into the Gulf a sinuous western coastline.

Predominantly northerly and northeasterly winds in the Gulf of Elat create a chimney effect resulting in the southward transport of sand (Friedman, 1968). Wind speeds are sufficient to create waves which strike

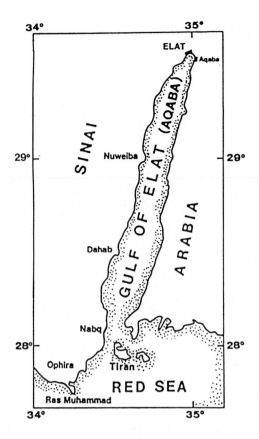

Fig.1. Location map of the Gulf of Elat (modified after Gavish, Krumbien, and Halevy, 1985, p. 187).

the Gulf's shoreline diagonally giving rise to substantial longshore currents. Currents may redistribute terrigenous debris brought into the Gulf by wadis and sweep some of the fine terrigenous and pelagic sediments from the submarine slopes into the deeps.

A typical example of fans and their associated coastal supratidal sabkhas is illustrated by the area of the mouth of Wadi Watir (Fig. 2). The Wadi Watir is one large trunk stream emerging from the edge of the mountains and spreading out into a large fan covering light sands and gravels. Composed of darker-colored and finer-grained sediment several small fans have grown across parts of the large fan (Fig. 3). Incised channels radiate across the surface of the large fan and dunes have spread across parts of its distal side with crests striking approximately east/west.

80

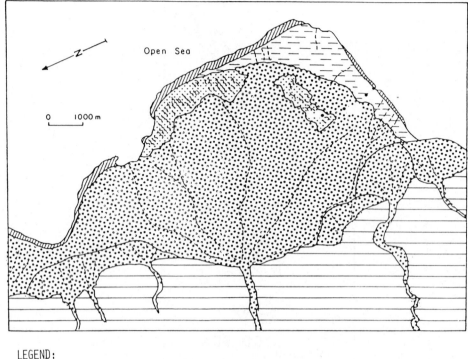

LEGEND:

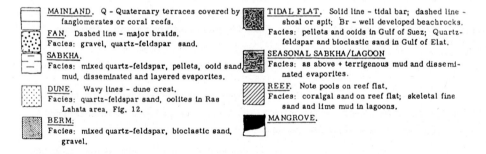

MAINLAND. Q - Quaternary terraces covered by
  fanglomerates or coral reefs.
FAN. Dashed line - major braids.
Facies: gravel, quartz-feldspar sand.
SABKHA.
Facies: mixed quartz-feldspar, pellets, ooid sand,
  mud, disseminated and layered evaporites.
DUNE. Wavy lines - dune crest.
Facies: quartz-feldspar sand, oolites in Ras
  Lahata area, Fig. 12.
BERM:
Facies: mixed quartz-feldspar, bioclastic sand,
  gravel.

TIDAL FLAT. Solid line - tidal bar; dashed line -
  shoal or spit; Br - well developed beachrocks.
Facies: pellets and ooids in Gulf of Suez; Quartz-
  feldspar and bioclastic sand in Gulf of Elat.
SEASONAL SABKHA/LAGOON
Facies: as above + terrigenous mud and dissemi-
  nated evaporites.
REEF. Note pools on reef flat.
Facies: coralgal sand on reef flat; skeletal fine
  sand and lime mud in lagoons.
MANGROVE.

Fig. 2. Sea-marginal environments in the area of Wadi Watir, Gulf of Elat.

A considerable volume of the terrigenous sediment being supplied to the
Gulf is from adjacent uplifted lands. Sediments eroded from the Precambrian
Arabo-Nubian massif in the southern part of the Sinai peninsula are composed
of igneous and metamorphic granitic rocks of the Upper Gattaria series and

Fig. 3. Oblique areal photograph of fans and sabkahs of Wadi Watir area,
Gulf of Elat. View due east showing bedrock of Arabo-Nubian Shield, fault
scarp, and fans of several sizes. One large fan, that of Wadi Watir,
underlain by light-colored sand and gravel. Several small fans composed of
darker-colored and coarser-grained sediment have grown across parts of the
large fan.

volcanic rocks including andesitic flows, tuffs, agglomerates, and
serpentines of the Dokhan series. Sediment from the southernmost tip of the
Sinai is supplied to the Gulf by the lime-grainstones of the Quaternary
Khashabi Formation (Bentor et al., 1974). On the southeastern side of the
Gulf a large aerial extent of Miocene evaporites and marine sediments supply
clastic debris (Brankamp, et al., 1963). Wadis draining through sedimentary
terrain in the central and northern Sinai to the Gulf is carried almost
exclusively sediment derived from Paleozoic-Mesozoic Nubian Sandstone
(Weissbrod, 1969).

There are second cycle gravels and sands being deposited on the alluvial
fans. Deposits of alluvium which now lie above sea level are comprised of
some sediments which accumulated about 4000 years B.P. when sea level
worldwide was 1 to 3 m higher (Fairbridge, 1961; Jelgersma, 1971; Stoddart,
1971) and of other deposits which have been tectonically uplifted. Along the
southern Gulf, coral reefs at several levels are also being dissected by
erosion and reflect similar sea-level changes and/or tectonic uplift.

The intertidal zone in the Gulf of Elat is relatively narrow. The uppermost intertidal zone is covered with boulders, cobbles and gravel cemented into beachrocks. At the mouths of wadis the intertidal flats consist of well-sorted sands, and beachrocks are extensively developed. Coarser-grained and poorly-sorted deposits are characteristic of the berm-top sediments. Channel mouths, initially connected to the sea, are blocked by berms trapping shallow, elongate pools in which terrigenous, organic-rich mud settles out. These back-berm channels quickly become hypersaline when flooded by marine water during periodic storms. More fresh water and terrigenous mud are also provided by flash-floods. The berm on the open sea side of the barrier is characterized by coral and mollusc fragments of cobble size and by igneous pebbles. Sandy supratidal flats are covered by oyster-reef mounds, seagrass, well-sorted sands comprising low tidal bars and by pseudo-oncolites (Fig. 4). The "oncolites" are several centimeters in

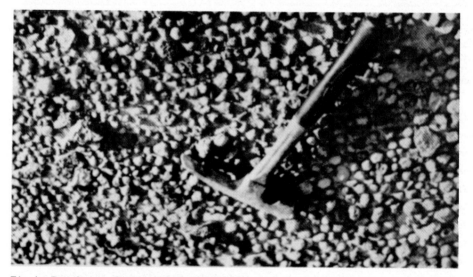

Fig.4. Pseudo-oncolites on intertidal flats, Gulf of Elat.

size, consisting of fine and medium-sized skeletal particles. Their internal structure shows one or two discontinuous, concentric, carbonate laminae. Rounded oncolites were found rolling on the intertidal flats and in the uppermost subtidal zone. Some column-shaped oncolites were found embedded in

the indurated host sediments. The rounded oncolites are probably products of
the disintegration of the column-shaped ones.

The distal margin of the fan of Wadi Watir, like that of many other fans,
is fringed with reefs. These reefs protect the sabkhas despite the tendency
of the southwest moving waves and currents to destroy them (Friedman and
Sanders, 1978, p. 302-304), allowing the close juxtaposition of flourishing
reef growth and copious terrigenous sediment deposition (Figs. 5 and 6). Fan
deposition in the Gulf of Elat does not seem to adversely effect the
organisms responsible for carbonate buildup. Almost the entire coastline of
the Gulf of Elat is fringed with corals and other reef-growing organisms.
Due to the steepness of the submarine slopes the reefs are restricted to
areas within 10 to 200 m of shore where the waters are shallow and are also
restricted to 10 to 100 m in width.

Prior to the reef a lagoon may exist whose width varies and profile
becomes steeper where it is narrower. Abundant near the lagoon strand line
and becoming progressively less common towards the reef are rounded pebbles
of Precambrian igneous or metamorphic origin from the nearby fans. Brown
algae, Padina pavonica Thivy, deposit an external skeleton of calcium
carbonate and are dominant only where the water is very shallow
(approximately 40 cm below low tide). Also common in the shallow water of
the lagoon is the black sea urchin Diadema saxatile. As the water gets
progressively deeper the brown algae becomes less common and sparse coral
patches begin to develop.

Boulder-sized rubble of dead coral, in blocks 1 to 2 m in diameter can be
found close to the reef. The mechanism of transportation of the
boulder-sized blocks 1 to 2 m in diameter and the coral rubble has been
attributed to momentary currents of high velocity which develop when storm
waves break on the reefs. Documentation of the formation of coarse reef
rubble in this fashion comes from the Australian Barrier Reef (Fairbridge
and Teichert, 1948) and the Jaluit Atoll in the Pacific Marshall Islands
(McKee, 1959). It was also observed during the 1969 Hurricane Donna on the
Florida Reef Tract. Winds attained a speed of 230 k/hr and tides were about
4 1/2 m above normal spring high tides. The storm-generated currents
succeeded in moving large blocks of dead coral (Ball, et al., 1967). The
Gulf of Elat boulders are coated with coralline algae and also support live
coral groups, indicating that the last great storm in the Gulf of Elat
occurred a long time ago.

Fig. 5. View vertically downward from an airplane of fans of several sizes and ages at the shore of the Red Sea. Bedrock (at top) consists of Precambrian gneisses composing Arabo-Nubian Shield. Seaward side of large fan underlain by light-colored sediment is fringed by coral reef (scarcely visible in this view). Dry bed of incised straight channel (about 2km long), which crosses fan from apex to periphery, displays well-developed pattern created by a braided stream. Drowning of seaward end of this dry channel has created small embayment in periphery of fan. Radiating network of faintly incised channels covers surface of large fan. Several small fans (upper left and lower right) have grown across parts of large fan. Waves approaching from lower right have built pointed spit on left side of large fan.

Progressively reefward, patch reefs begin to develop which ultimately merge with the reef core. Near the reef core, large (15 to 20 cm long) broken coral fragments are abundant. Contigious with, and derived from the reef, are thick blankets of skeletal sand which occur on both lagoonal and seaward sides of the reef. Skeletal remains of foraminifers, molluscs, and echinoderms which live and die in and around the reef community are important constituents of reef-derived sands. Parrot and trigger fish are among the many genera of fish that chew on the reef producing skeletal sand

85

Fig. 6a. Fans have built reefs, Ras Abu Gallum, west side of Gulf of Elat, viewed obliquely from the air. Distant view showing lobe of fan in foreground that has built completely across the reef.

Fig. 6b. Fans have built reefs, Ras Abu Gallum, west side of Gulf of Elat, viewed obliquely from the air. Closer view of smaller lobe of fan (at top of 6a) that has been modified by waves, which have deposited a series of beach ridges (lower left).

(Friedman, 1968). At a depth of 15 m and a distance of 40 m from the main reef, this sand is dark-colored, a possible result of reducing conditions at and below the water-sediment interface at that depth.

Skeletal sands persist all the way to the bottom of the Gulf, but where the water is deep they contain abundant admixtures of terrigenous debris derived from the Precambrian crystalline rock and ancient carbonate reef terraces and deposited in the Gulf by flash-floods from the wadis and by wind from the Arava Valley in the north. The terrigenous constituents are reflected in the amounts of insoluble residue and in their chemical makeup. A deep-sea core (v. 14-126) collected by the Columbia University R/V Vema in 1960 contains carbonate material ranging from 20 to 75% of the total. Analysis of the mineral content and chemical makeup (especially their high-magnesium content) made it possible to determine that coral fragments predominate over molluscan fragments (Friedman, 1968, p. 944). Coralline algae synthesize high-magnesium carbonate and are major contributors to the reef-derived sands. Size-frequency analysis of the reef-derived sands showed them to be less well-sorted than nearby carbonate beach sands and the fines (<62μm to <125μm) had been washed out of the carbonate beach sands, presumably by wave action. Rapid transport of the skeletal sand is indicated, otherwise organisms would have broken down the sand into mud. Aragonite is sporadic or absent and the unusual combination of high-magnesium and low-magnesium calcite was considered to be attributable to the faunal assemblage (Sanders and Friedman, 1967).

The Gulf of Elat marginal-slope deposits were considered to be composed of coarsely stratified deposits (Ben Avraham, et al., 1977). However, in 1977 the Woods Hole Oceanographic Institute's R/V Atlantis II took five deep-sea cores from marginal slopes along the western side of the Gulf and found two types of sediments: (1) grey-green partly silty calcareous mud and (2) silty, clayey sands, mostly quartzose and occasionally arkosic. In the northern Gulf, two cores taken at depths of 550 and 858 m adjacent to the Elat Deep, contained turbidites of alternating silty muds and quartzose sands while in the southern Gulf, a core taken at a depth of 963 m contained a "thick arkosic sequence in the lower part of the core" (Reiss, et al., 1980, p. 295).

The deposition of terrigenous clastic debris side by side with the accumulation of a significant buildup of carbonates is an unusual aspect of the submarine sedimentation in the Gulf of Elat area. Factors suggested for this carbonate/terrigenous occurrence are: 1) the coarseness of the terrigenous debris makes the detritus less of a problem for the reef-building organisms (Friedman, 1982); 2) turbidity currents in the submarine canyons may carry the bulk of the fine materials washed down the wadis directly into the deep waters where the silt and mud cannot effect the reef builders; 3) the survival of the reefs is favored by the long periods of quiesence between flash-floods.

Lateral transitions between coexisting carbonate and terrigenous environments may result in a mixed sediment observable in vertical section. Sediment produced by "facies mixing" (Mount, 1984) are rare since the contact between the contrasting lithofacies is usually sharp due to: (1) abrupt later transitions where mixed sediments are unlikely to be preserved, (2) a fundamental alteration in depositional conditions such as erosion or rapid migration of environments (Mount, 1984). Environments where this type of mixed sediment occurs include tidal flats, coastal dunes, carbonate shoal complexes, and reef tracts.

2.2 Java Sea Pulau Seribu Reefs

An example of facies mixing carbonate buildups in a principally clastic environment exists in the Java Sea where carbonate reefs thrive despite the deposition of clastic sediment into the sea.

The Pulau Seribu Reef group is a chain of carbonate islands stretching across for 40 km and located 25 km north from the island of Java in the Java Sea (Fig. 7). The reefs grow on sea floor structural highs trending NNE-SSW, separating the Sunda sub-basin to the west and the Arjuna sub-basin to the east. Each individual reef varies from a few meters to more than a kilometer in width and tends to be roughly elongated in an east-west direction.

Scrutton (1976) has given an excellent description of the carbonate facies distribution of the Pulau Seribu Group (Fig. 8). He recognizes the sediments on the reefs are chiefly skeletal sands and gravels. The main skeletal components are corals, molluscs, and Halimeda, with minor components of echinoids, foraminifera, and red algae. Small vegetated cays are present in most of the reefs and consist of wave and wind accumulated skeletal sand deposits. Carbonate mud is restricted to lagoons which may exist between the sand cays and the reef flats. Reef flats are the result of the formation of skeletal sands and gravels by the destructive process of wave action on young reefs. The main generation zone for reef growth is the reef edge or "growing edge". The optimum conditions for growth occur on the east and west sides of the reef, there fauna attains maximum abundance and diversity. The drop from the growing edge to the sea floor is dramatic with slopes an average of 70-75°. Coral diversity diminishes gradually with depth

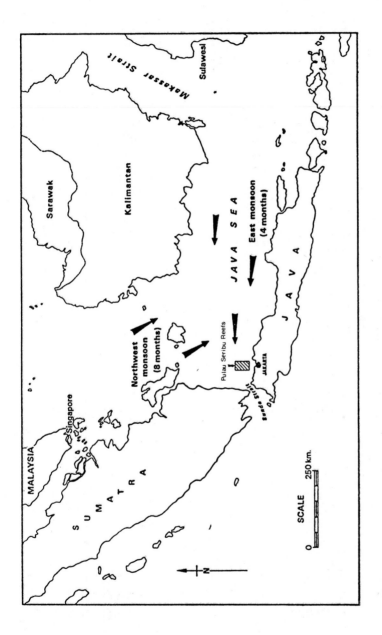

Fig. 7. Location map of Java Sea Pulau Seribu Reefs (modified after Sctutton, 1976, p. 26).

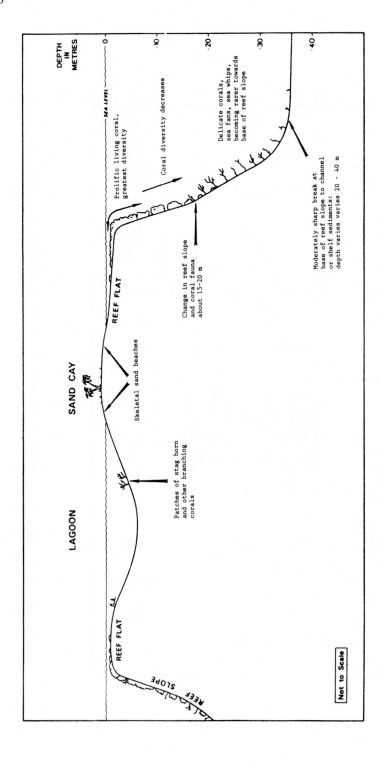

Fig. 8. Diagrammatic section through a typical reef (Scrutton, 1976, p. 28).

and becomes irregular patches. Narrow channels separate the individual reefs and may become up to 3 km wide and approximately 35-60 m deep. The flat bottoms contain sediments with relatively little reefal debris and up to 60% clay and silt.

Three major sedimentary facies have been described (Scrutton, 1984): (1) the Coral-molluscan facies (reef top), (2) the Coral-Halimeda facies (reef flank), (3) the shelf facies, and a transitional facies.

The Coral-molluscan facies, found on the reef top, includes the sand cay, reef flat and lagoon physiographic zones. It is characterized by well-sorted coral dominated skeletal sands with abraded grains indicative of a high energy environment. Coral again dominates the Coral-Halimeda facies on the reef flank, but it is a poorly-sorted facies with a notable absence of abrasion. The contact between these two facies is sharp with very little occurrence of sediment mixing. Mixing of sediment begins in the shelf facies which contains silty clays with abundant shelf debris. The base of the reef slope marks the transition to shelf sediments as the ratio of clastic sediments to carbonate material increases resulting in facies mixing. Also considered a transitional facies are the wide inter-reef channels containing carbonate-clastic sediment mixtures.

The most influential control on reef growth and morphology appears to be the seasonal current and wind directions. Monsoons originate from an easterly direction for 4 months out of the year and from a westerly direction for 8 months resulting in long-shore currents which transport the fine-grained material laterally away from the reef environment onto the substratum of the Java Sea. Therefore 25 km away from the island source of this terrigenous material, the water is clear, allowing active coral reef growth. These long-shore currents account for the east-west elongated nature of the reefs. In order to obtain maximum benefits from the current-supplied nutrients, living corals grow preferentially towards the prevailing current explaining the heavy coral growth on the east and west sides of the reefs (Fig. 9).

The varying current and wind directions inhibit the development of large reef complexes with well-defined fore-reef, reef, and back-reef environments. Without these strong directional currents, however, carbonate-secreting organisms may not exist as prolifically as they do. High rainfall in this equatorial climate ensures a very substantial sediment-ladened run-off from the islands of Sumatra, southern Kalimantan, and Java (Fig. 10). This clastic sediment, chiefly clay with small amounts of silt contains angular quartz, glauconite, and rock fragments, is

92

Figure 9.   Pulau Seribu reefs elongated east-west reflecting the direction
of longshore currents, Java Sea.

Figure 10.   Sediment-ladened river depositing its load into the Java Sea.

detrimental to the growth of reef complexes. Its redistribution by the longshore currents and accumulation on the substratum of most of the Java Sea is important for reef development. Sea floor highs are the only suitable areas for reef developments where shallow-sea currents winnow out the fine-grained material.

SUBMODERN ANALOGUE

The Yeroham Basin, Southern Israel: Neogene carbonate-clastic sequences

A submodern example of in situ mixing can be found in the upward-coarsening Crassostrea sequences of Neogene Basins in southern Israel where bioherms and Miocene beachrock are an integral part of a fine-grained terrigenous section. An intermittent stream (wadi) dissecting the rocks of the Hazeva Formation in the Yeroham Basin (Fig. 11) exposes a vertical sequence displaying at least seven cycles of sedimentation (Fig. 12). Each cycle is composed of a regressive clastic coarsening-upward sequence with transgressive events separating the cycles. Beginning with siltstone, a cycle grades upward into very fine to fine-grained sandstone. At the top of each cycle there is an erosion-resistant, coarse-grained facies such as sandstone, beachrock or oyster bank, indicative of a shoaling phase. These coarsening-upward sequences are the result of regressive sedimentation in a transitional environment such as a sea margin where terrigenous sediments may accumulate. The oyster bank topping a prominent elliptical ridge in the erosional basin provides information on the environment in which the carbonate and terrigenous sediments were deposited.

The mutually exclusive relationship between fine-grained terrigenous sediments and carbonate-secreting organisms is not necessarily the case with oysters, particularly the crassostreids as supported by the presence of oyster mounds and reefs in estuaries, deltaic lobes and other coastal bodies since Cretaceous time (Friedman and Sanders, 1978). In the Yeroham Basin the homogeneous mass of oysters are in growth position. While borings from gastropods and sponges have been recorded, no large-scale accumulation of skeletal debris was observed. Therefore, in situ growth apparently outstripped predation. Characteristic of the reef base section are 'ghosts' of young mollusks, highly resistant to the post-depositional dissolution by percolating meteoric waters that has overtaken the mature fauna. Present are pelecypod ghosts, calcite cement and travertine, especially at the base of the bank, indicating large-scale recycling of calcium carbonate.

By analogy with modern environments, the Yeroham oyster bank is considered to have been intertidal. It was exposed subaerially at ebb tide in a deltaic setting similar to that observed at the mouth of the

94

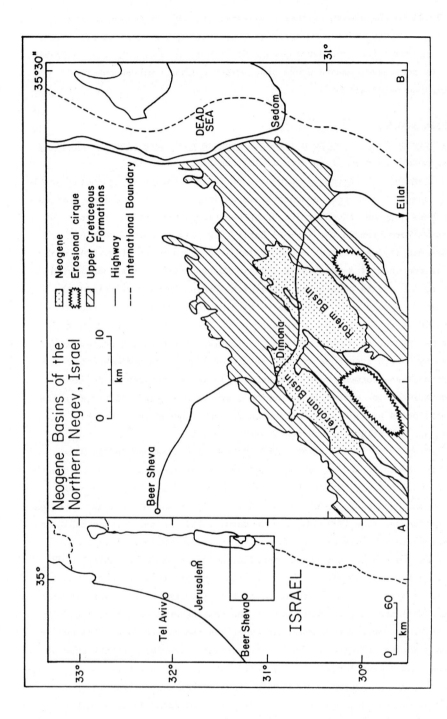

Fig. 11. Location map of the Yeroham Basin, southern Israel.

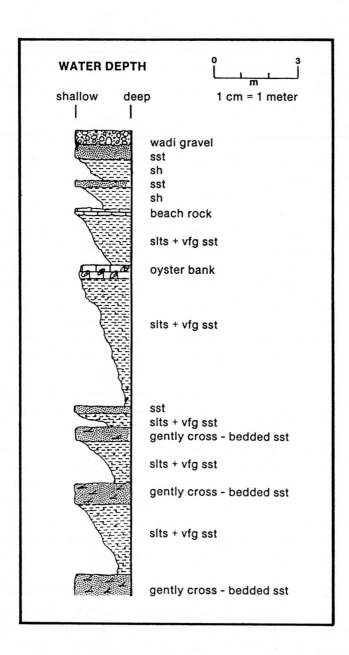

Figure 12.   Base and seven upward-shoaling cycles in Yeroham Basin. Oyster reef tops cycle 4.

Atchafalaya River in Louisiana where widespread reefs occur where the river dumps approximately half a million tons per day of fine-grained clay particles into the Gulf of Mexico. Oysters have adapted to life in the muddy and turbulent waters by eliminating clay-mucous particles as pseudo-feces (Bahr and Lanier, 1981) which may have been a paleoecological retreat into an environment where abundant nutrients are washed in from the land and the oysters' predators cannot survive.

CONCLUSION

In attempts to reconstruct ancient depositional settings for the purpose of locating new carbonate petroleum reservoirs, studies of modern carbonate depositional environments have led to the observation that carbonate-secreting organisms and terrigeneous sediments may coexist. The three case histories discussed exemplify three very different settings in which clastics are contemporaneously deposited with carbonates:

1. In the Gulf of Elat, rare, high-intensity sedimentation events as well as in situ mixing and erosion of exposed carbonate source terrains are the processes which result in the coexistence of terrigenous sediments and reef complexes. The hot, dry climate affects the sedimentary processes with respect to the amount and frequency of sediment run-off, and the development of evaporites, algal-mat carbonates and fringing coral reefs. Longshore currents created by northerly and northeasterly winds remove the fine-grained sediments away from the submarine slopes into the deep basin. The coarseness of the remaining sediment and the infrequency of terrigenous deposition allows for the development of carbonates on fans and the growth of fringing reef complexes.

2. The Java Sea area is characterized by facies mixing of fine-grained terrigenous sediments and carbonate-secreting organisms. The humid climate results in a consistently high run-off of fine-grained sediments. The seasonal winds and currents from the monsoons create longshore currents which laterally remove the fine muds and clays away from reef complexes which thrive in clear water approximately 25 km from the Java coast.

3. In situ mixing of a different type occurs in the Neogene Yeroham Basin in southern Israel. Carbonate-secreting organisms (oysters) actually prefer a sediment-ladened environment. In what may have been a paleoecological retreat into a setting where predators could not survive, oysters (Crassostreids) adapted to life in intertidal, muddy waters by eliminating fine sediments as pseudo-fecies.

REFERENCES

Bahr, L.M. and W.P. Lanier, 1981. The ecology of intertidal oyster reefs of the South Atlantic coast: A community profile. U.S. Fish and Wildlife Services, Washington, D.C., FWS/OBS 81/15, p. 105.

Ball, M.M., Shinn, E.A. and Stockman, K.W., 1967. The geologic effects of Hurricane Donna in South Florida. Jour. Geology, v. 75, pp. 583-597.

Ben-Avaraham, Zvi, Almagor, Gideon and Garfunkel, Zvi, 1979. Sediments and Structure of the Gulf of Elat (Aqaba) - Northern Red Sea. Sedimentary Geol., v. 23, pp. 239-267.

Bentor, Y.K., Bogoch, R., Eyal, M., Garfunkel, Z. and A. Shimron, 1974. Geological map of Sinai, 1:100,000, Jebel Sabbagh sheet.

Brankamp, R.A., Brown, G.F., Holm, D.A. and Layne, N.M., Jr., 1963. Geology of the Wadi As Sirhan Qudrangle, Kingdom of Saudi Arabia. U.S. Geol. Survey, Miscs. Geol. Investigations Map 1-200A.

Fairbridge, R.W., 1961. Eustatic changes in sea level. In: Physics and chemistry of the earth, v. 4: New York, Pergamon Press, pp. 99-185.

Fairbridge, R.W. and Teichert, G., 1948. The low isles of the Great Barrier Reef: a new analysis. Geogr. Jour., v. 3, pp. 67-88.

Friedman, G.M., 1968. Geology and geochemistry of reefs, carbonate sediments, and waters, Gulf of Aqaba (Elat), Red Sea. Jour. Sed. Petrology, v. 38, no. 3, pp. 895-919.

Friedman, G.M. and Sanders, J.E., 1978. Principles of Sedimentology,. New York, John Wiley & Sons, p. 792.

Jelgersma, S., 1971. Sea-level changes during the last 10,000 years, In: Steers, J.P., (Ed.),. Introduction to coastline development. London, Macmillan Co., pp. 25-48.

Lebedev, A.N., (Eds.), 1970. The climate of Africa, Part 1: Air temperatures, precipitation (trans. from Russian). Jerusalum, Israel Program for Scientific Translations, p. 482.

McKee, E.D., 1959. Storm sediments on a Pacific atoll. Jour. Sedimentary Petrology, v. 29, pp. 354-364.

Mount, F., 1984. Mixing of siliclastic an carbonate sediments in shallow shelf environments. Geology, v. 12, pp. 432-435.

Reiss, Z., Luz, B., Almoqi-Labin, A., Halicz, E., Winter, A. and Wolf, M., 1980. Late Quaternary paleoceanography of the Gulf of Aqaba (Elat), Red Sea, Quaternary Res. 14, pp. 294-308.

Scrutton, M. E., 1976. Modern Reefs in the West Java Sea, In: Proceedings of the Carbonate Seminar, Jakarta, 1976; Indonesian Petroleum Association, Special Volume, pp. 14-36.

Stoddart, D.R., 1971. Geology and morphology of reefs, In: D.R. Stoddart (Ed.), Regional variations in Indian Ocean coral reefs. London, Academic Press, pp. 3-38.

Weissbrod, T., 1969. The Paleozoic of Israel and adjacent countries. Geol. Survey Israel Bull. 48, p. 32.

Chapter 4

GULFS OF THE NORTHERN RED SEA: DEPOSITIONAL SETTINGS OF ABRUPT
SILICICLASTIC-CARBONATE TRANSITIONS

H.H.ROBERTS and S.P. MURRAY
Coastal Studies Institute, Louisiana State University, Baton Rouge LA   70803

ABSTRACT
    The two narrow gulfs of the northern Red Sea, Gulf of Suez and Gulf of
Aqaba, have had different tectonic histories, but both display active
interfingering of siliciclastic and carbonate facies. In an early stage of
rifting, these embryonic seas are flanked by rugged mountains (to 2,000 m) and
narrow coastal plains (generally <10 km wide) built of alluvial fans composed
of poorly sorted siliciclastic debris, mostly from crystalline basement rocks.
An arid setting promotes aperiodic transport of siliciclastic sediments as well
as deposition of evaporites (coastal sabkhas) and carbonates (reefs and
associated sediments). Gulf margins prograde by a combination of rapid fan
deposition during flash floods and subsequent carbonate stabilization of
terrigenous fans and cones during intervening periods. High-resolution seismic
and side-scan sonar data suggest that narrow pathways for sediment transport
through the reefs are continually active. They accommodate most of the sediment
transport to deep water during small discharge events. Large flash floods may
completely overwash carbonates at the distal ends of fans, requiring renewed
reef development. Rapid siliciclastic deposition, coupled with biological and
chemical binding of carbonates as well as their tendency toward vertical
buildups, results in steep slopes along the gulf margins.
    The Gulf of Suez is shallow (<100 m), and a relatively broad (>12 km) and
geometrically complicated strait separates it from the northern Red Sea. In
contrast, the Gulf of Aqaba is deep (<1,800 m) and has a very narrow strait.
Although both basins result from rifting associated with opening of the Red
Sea, the Gulf of Suez is dominated by normal faults and tilted blocks, whereas
the Gulf of Aqaba formed primarily by strike-slip displacements, with minor
movements perpendicular to its extension. Seismic and borehole data confirm
that the Gulf of Suez is a grabenlike structure that has filled with nearly 6
km of dominantly siliciclastic sediment since Miocene times. An evaporite unit
over 1 km thick and numerous thin carbonate horizons, as well as local reef
buildups, interfinger with the noncarbonates. New data suggest that local
basins within the Gulf of Aqaba have as much fill as the Gulf of Suez.
Turbidites and foraminifer muds presently are filling the deepest basins.
    The Gulf of Suez contains numerous carbonate platforms, which are probably
seated on subtle gulf-parallel structures. Some of these features suggest that
they are the initial stages of much larger carbonate platforms that will
develop as rifting continues. Modern physical processes--strong axial winds
(<30 m/sec), an energetic gulf-parallel wave field, and vigorous tidal currents
(>50 cm/sec)--tend to streamline reefs and sediment bodies, creating spindle
shaped carbonate platforms. The Gulf of Aqaba has no mid-gulf platforms, but a
complex of reef-dominated carbonates exists on gulf-normal structural blocks at
the Strait of Tiran. A cross-section reduction of this already narrow strait by
lowering of sea level, reef growth, and/or sedimentation could drastically
change the basin-filling process by increasing salinity to the point of
eventual evaporite deposition.

1 INTRODUCTION

Coexistence of well-defined siliciclastic and carbonate facies is known from the study of both recent and ancient depositional systems. In most cases, however, the approach has been to perform detailed investigations on either the siliciclastics or carbonates and focus little attention on the transitions between these distinctly different sedimentary systems. A good example of how emphasis on either carbonates or siliciclastics within a basin can bias our thinking about depositional systems was recently given by Purser (1982) concerning the Arabo-Persian Gulf. Although the "Persian Gulf" has long been considered a classical model for modern carbonate sedimentation, this sedimentary basin is dominated by terrigenous deposition. Even the carbonate province of the Persian Gulf, which occurs along the tectonically stable Arabian side, is impacted by eolian siliciclastics, which can be dominant in local areas (Shinn, 1973).

With the increasing need for a detailed understanding of sedimentary settings and their facies associations that can be used by industry to predict lithologic variability, especially in hydrocarbon exploration, it is important that more attention be focused on siliciclastic-carbonate transitions. Investigation of present-day sedimentary basins can provide critical information about local and regional processes that influence styles of deposition for both siliciclastics and carbonates. Recent studies are singularly important in sedimentology because they allow us to actually monitor processes that control facies development. Such processes help to determine the spatial and temporal scales of deposition within any given regional geologic framework. This paper addresses the manner in which both siliciclastics and carbonates are deposited in the relatively small, tectonically controlled gulfs of the northern Red Sea. It is the objective of this work to summarize both sedimentological and physical process data from the gulfs of Suez and Aqaba in order to help organize and expand our understanding of sedimentation in fault-controlled marine basins developing under arid climatic conditions, systems of abrupt siliciclastic/carbonate transitions.

2 DATA SOURCES AND METHODS

Primary data on which this paper is based were generated during two field projects designed to study the marine geology and physical oceanography of straits between the small gulfs of Suez and Aqaba and the Red Sea. The first study was conducted in the southern Gulf of Suez, Jubal Strait area, during June-July 1981. A similar project was fielded in February-March 1982 at the entrance to the Gulf of Aqaba (Strait of Tiran) (Fig. 1). The Gulf of Suez

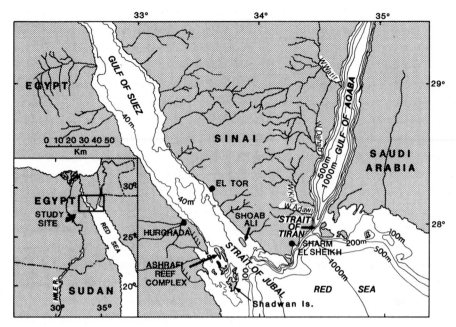

Figure 1. Location map of the northern Red Sea area showing the tectonically controlled Gulf of Suez and Gulf of Aqaba.

study was conducted from a 22-m fishing boat, the Karam El Suez (Port Suez), adapted for research purposes. In the Tiran Strait investigation, the R. V. Shikmona (Haifa), a research vessel from the Israel Oceanographic and Limnological Research Institute, was utilized in a cooperative scientific effort with Institute and Israeli Navy personnel. Hurghada (Fig. 1), the site of the Egyptian Academy of Sciences' Red Sea Laboratory, served as base of operations for the Strait of Jubal work. Sharm el Sheikh (Fig. 1), near the southern end of the Sinai, was the home port and staging area for the Tiran Strait study.

The same basic data collection approach was used in each investigation. However, the Strait of Tiran project was more equipment intensive. Endeco 174 recording current meters were used to monitor the shallow flow field (<100 m). Six of these meters were deployed across the Strait of Jubal. Because of deeper passages in the Tiran Strait, Aanderas RCM4 meters were deployed near the bottom in the Enterprise and Grafton passages (Fig. 1). The Endeco meters recorded current speed and direction, conductivity, and temperature. Aanderas meters recorded current velocity and temperature. STD casts were made in both study areas with a Guildline Model 8776 digital recording conductivity-

temperature profiling system. Both projects deployed recording anemometers
(Weather-Measure Skyvane I) to continuously monitor wind speed and direction.
Pressure sensors were installed at the bases of operation (Hurghada and Sharm
el Sheikh) as well as in the field study area to record tide levels.

Marine geology data were collected using a variety of methods and
instrumentation. Bottom sampling, side-scan sonar, high-resolution seismic,
echo sounder profiling, and SCUBA for direct observation/sampling provided the
data base. With exception of seismic profiles, which were not collected in the
Gulf of Suez, data sets were similar for the two study areas. An EG and G
Uniboom seismic source (Model 230-a) was used with a Model 232 capacitor
bank-power supply and a Model 263-B hydrophone array to acquire high-resolution
seismic data from the Tiran Strait area. These data were recorded on an EPC
4600 graphic recorder. Precision depth recordings were made with an EDO-Western
12-kilocycle transducer and a Model GDR-1C-19T Gifft recorder. A Raytheon Model
231 echo sounder was used in the Jubal Strait area. A Klein side-scan sonar
system provided excellent images of the sea floor. All positions of
instrumentation, samples, and survey lines were located using an electronic
range-range locating system (Decca Del Norte Trisponder). In collecting
side-scan sonar, high-resolution seismic, and echo-sounder survey data, all
recorders were linked to the Trisponder system, and position updates were
recorded at 1-minute intervals.

3 TECTONIC AND SEDIMENTOLOGIC FRAMEWORK

Both the Gulf of Suez and its neighbor, the Gulf of Aqaba, are products of
Tertiary rifting associated with opening of the Red Sea. They are basins of
deposition that display rapid transitions from rugged flanking topography to
the basin floor (distance of <20 km). However, these two narrow basins are
different in their present bathymetric configuration. The Gulf of Suez is
shallow and filled with sediments and carbonate buildups (average <70 m),
whereas Ben-Avraham et al (1979) indicate that the Gulf of Aqaba reaches depths
as great as 1,850 m and is segmented into subbasins, with reefs occurring
dominantly along the basin margins (Fig. 1).

Structural origins of these two embryonic seaways are somewhat different.
Garfunkel and Bartov (1977) suggest that structural fabric of the Suez Gulf
cuts through the relatively stable margins of the Arabo-Nubian Shield and is
dominated by NW-SE trending normal faults devoid of any compressional
structures. In contrast, the Gulf of Aqaba displays evidence of transform
movement, which has allowed widening of the Red Sea relative to the Gulf of
Suez. Mart and Hall (1984) show that the transition zones from the Red Sea to

these gulfs are also significantly different. Ben-Avraham (1979) and Garfunkel
(1981) confirm the Dead Sea rift of which the Gulf of Aqaba occupies the
southern part, to be a plate boundary of the transform type (partially leaky).
This system of transform as well as normal faults connects the Red Sea's area
of seafloor spreading with the Zagros-Taurus zone of continental collision.
Tectonic heritage of both these gulfs has led to mountainous margins consisting
primarily of tilted blocks of basement rock, which rise to typical elevations
of ~1,000 m above sea level and maximum elevations of ~2,000 m. Erosion of
these igneous and metamorphic rocks produces sediment that has led to the
construction of narrow coastal plains bordering each gulf.

Because the Gulf of Suez is an important area for hydrocarbon production and
exploration (Gilboa and Cohen, 1979; Kanes and Abdine, 1983), considerably more
is known about its sedimentary fill and structure as compared to the Gulf of
Aqaba. Both seismic and drilling data confirm a sedimentary fill in the Suez
Gulf in the range of 4 km to more than 5 km thick. Garfunkel and Bartov (1977)
indicate that block faulting responsible for the graben structure of the Gulf
was underway by Oligocene-Early Miocene. Erosion of topographically high
margins provided the initial fill. Ben-Avraham (1984) indicates that sediments
contained within two of the three subbasins identified in the Gulf of Aqaba are
about the same thickness as sedimentary fill in the Gulf of Suez, ~5 km.

Figure 2 illustrates surface geology of the southern Gulf of Suez, plus an
interpretive cross section based on data collected from a transect of oil
across the Suez graben. Thickest sedimentary sections (~5 km) occur in the
western basinal structures, while the most compressed fill sequence (~1.5 km)
lies beneath Shoab Ali, in the southeastern Gulf (Fig. 1). Pre-Miocene
sediments (Fig. 2) consist of Paleozoic sandstone and upper Cretaceous to early
Tertiary clastics, carbonates, and evaporites resting on igneous and
metamorphic basement. Stratigraphic evidence suggests that within this complex
pre-Miocene unit, Lower Eocene strata show the first signs of movement of the
northwest-trending Gulf of Suez rift (Glen Steen, GUPCO, 1981, personal
communication).

Miocene strata comprise most of the Gulf of Suez sedimentary fill.
Deposition of these units was controlled by the fault-bounded gulf margins, the
Sinai Massif to the northeast and the Red Sea Hills to the southeast (Fig. 2).
Miocene deposition started with a terrigenous clastic sequence up to ~1.5 km
thick, showing both lateral and vertical variability in thickness and sediment
properties. Limestone and mudstone units are present, but are a minor part of
this early Miocene sequence. Widespread evaporites of early middle Miocene age
overlie the siliciclastics. These thick salts, with anhydrites and minor
interbeds of siliciclastics, limestones, and mudstones, are up to approximately

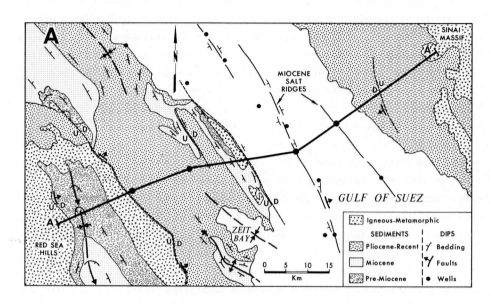

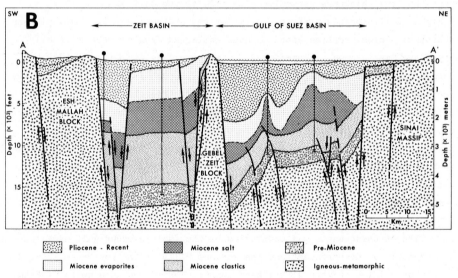

Fig. 2. A, Geologic map of the southern Gulf of Suez indicating major surface rock exposures, major structural features, and the positions of wells from which data were collected for constructing a geologic cross section. B, Cross section of the southern Gulf of Suez showing the fault-controlled structure of the basin and major subdivisions of its thick sedimentary fill. (Data were provided courtesy of Gulf of Suez Petroleum Company).

2 km thick, especially in the western basin. Evaporite units, described as
sabkha deposits by Abdine and Kanes (1985) have been diapirically and
structurally molded into elongate swells that have affected the thicknesses of
post-Miocene sedimentary units and locations of some reefs (e.g., Shoab Ali).

Pliocene-Recent deposition is up to ~2 km thick in the western parts of the
two grabens separated by the basement block, Gebel Zeit (Fig. 2). These
deposits represent the products of rejuvenated subsidence of the rift basin
floor relative to the basement rock flanks. Siliciclastics are dominant in the
initial stages of deposition. Reefs and other limestone deposits now overlie
the thick siliciclastics, suggesting a slowing of tectonic activity.

4 CLIMATE AND MARINE ENVIRONMENT

Both climate and geologic framework play active roles in the processes of
sedimentation that result in basin filling. They also determine many
characteristics of the dynamic marine environment. Gulfs of the northern Red
Sea have an arid climate, with rainfall ranging between 10 and 70 mm/yr (U.S.
Naval Oceanographic Office, 1965). As a product of these conditions,
evaporation rates in the gulfs of Suez and Aqaba are among the world's highest.
Estimates of evaporation range from 200 cm/yr (Neumann, 1952) to a more recent
and revised value of 365 cm/yr (Assaf and Kessler, 1976). As expected, the
highest rates of evaporation occur during the summer (Fig. 3).

Utilizing the Assaf-Kessler estimate of 365 cm/yr and a Gulf of Suez area of
8,000 km^2 (250 km by 32 km; Morcos, 1970), a value of 2.7 x 10^{10} m^3/yr can be
computed for water lost to evaporation. Moisture flux to the atmosphere
produces hypersaline (dense) waters in the gulf interiors, which can result in
bottom flow directed into the larger and less saline Red Sea. Such extensive
evaporation requires compensatory flow into the small gulfs to replace the
water mass lost to evaporation. Murray et al (1984) recently showed that
density-driven outflow and gulfward-directed surface exchange form a
well-developed two-layered flow system in the Strait of Tiran (Fig. 4), between
the Gulf of Aqaba and the Red Sea. Figure 5 illustrates this strong flow regime
from the deep (~300 m) Enterprise Channel of the Tiran Strait. Records from
both bottom and surface meters illustrate the strong semi-diurnal tidal
oscillation, with density-driven outflow and surface layer inflow velocities
averaging over 50 cm/sec.

A somewhat less distinct two-layered flow has been documented from the
comparatively shallow (<100 m) central and eastern Jubal Strait (Fig. 6),
southern Gulf of Suez (Murray and Babcock, 1982). In the Gulf of Suez, flow is

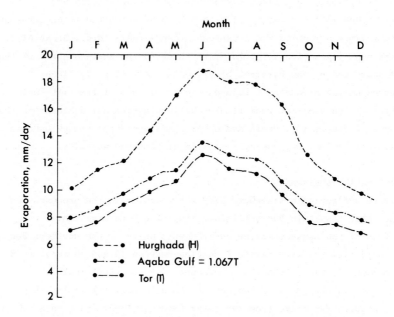

Figure 3. Monthly evaporation rates for sites in the Gulf of Suez and Gulf of Aqaba (Griffiths and Soliman, 1972).

bidirectional at the semi-diurnal tidal frequency. However, field studies have shown (Murray and Babcock, 1982) that net flow in the surface layer is directed gulfward, which is against the downwind direction. In contrast, the bottom layer of the Jubal Strait displays a mean flow directed south into the Red Sea. Along the strait's western margin, the combined effects of shallow water and complicated topography eliminate a two-layered flow system. In this part of the Gulf, persistent winds from the north cause a mean surface drift to the south.

Both the gulfs of Suez and Aqaba are narrow, elongate basins that have steep fault-controlled mountain fronts (to 2,000 m) paralleling their perimeters. Strong axial winds blow down those gulfs as a product of topographic steering of regional winds, augmented by diurnal heating and cooling of the basin-facing mountain slopes. The Gulf of Suez, in particular, experiences persistent strong winds of 8-15 m/sec from the north. Figure 7 illustrates wind speed and direction data collected from the southern Gulf (near Ashrafi, Fig. 6), where strong winds and accompanying waves consistently arrive from a northwesterly direction roughly parallel to the gulf axis.

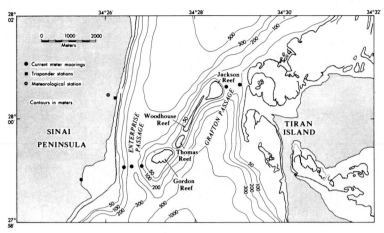

Figure 4. Map of the southern Gulf of Aqaba showing the reef-dominated Strait of Tiran. Note the two main passages, central reef complex, and the positions of current meter moorings within each.

Even though these small gulfs are fetch-limited systems, coincidence of local winds with the long axis of each elongate gulf produces unexpectedly strong waves. The maximum fetch of these water bodies is over 200 km. At the downwind end of this fetch, strong waves develop that are focused on the straits, their islands, and carbonate platforms.

Tidal range, as indicated by Morcos (1970), is ~0.3-1.5 m in the Gulf of Suez and ~0.3-0.75 m as measured at Sharem el Sheik, just south of the Tiran Strait, and entrance to the Gulf of Aqaba (Murray et al, 1984). Records from both areas display a well-developed semi-diurnal tidal signal. In the Gulf of Suez, tidal fluctuations describe a first-mode (one-half wavelength) standing wave. Tidal ranges decrease from the Jubal Strait to a nodal point near El Tor, about 60 km to the north (Fig. 1). Tidal ranges then increase again (~1.5 m) at the Gulf's northern end. These tidal oscillations are important in the Gulf because they drive strong tidal currents that influence sediment transport. Vercelli (1931), who published the first current observations in the Gulf of Suez, reported tidal current velocities of about 75 cm/sec at springs and 25 cm/sec at neaps along the central axis of the Gulf. In the Strait of Jubal, current velocities were reported to reach 100 cm/sec.

108

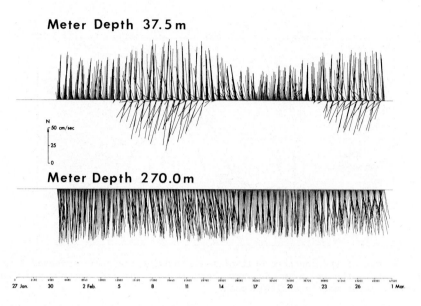

Figure 5. Observed data from two meters moored in the middle of Enterprise Passage, Tiran Strait. The top meter was moored at -37.5 m in the upper layer, which illustrates dominant flow into the Gulf. The bottom meter (moored at -270 m) monitored constant outflow of over 50 cm/sec into the Red Sea. Data were smoothed slightly with a 3-hour running mean.

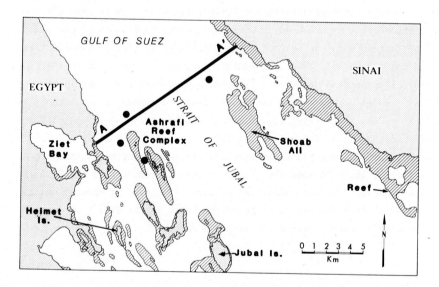

Figure 6. Map of the southern Gulf of Suez showing the Strait of Jubal, which is complicated by numerous islands, carbonate platforms, and reefs. Note the sites of current meter moorings (dots) and STD bottom sample transect, A-A'.

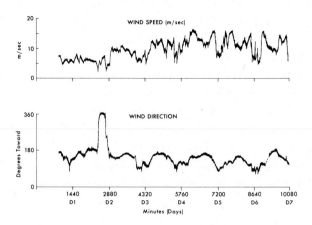

Fig. 7. Wind speed and direction data collected for a period of 7 days near the Ashrafi reef complex in the southern Gulf of Suez (Fig. 6). The sample interval was 10 minutes. Start time was 1457 h, 5 July 1981. Note the nearly unidirectional nature of the wind field and the abnormally high speeds during the period 8-11 July. The gulf axis is approximately 150°.

5 GULF MARGINS AND COASTAL ENVIRONMENTS

5.1 Alluvial Fans

Narrow coastal plains (generally <10 km wide) of the northern Red Sea gulfs are dominated by alluvial fans building directly into the marine environment. A combination of high relief near the basin, mechanical weathering in an arid climate, and the infrequent but effective flash flood as a sediment transport process has produced a belt of alluvial material that literally encircles the Red Sea and its young tectonic gulfs. Although Hayward (1985), Friedman (1982), Sneh (1979), Friedman and Sanders (1978), and Gvirtzmann and Buchbinder (1978) have discussed these common coastal features, comparatively little has been published about details of their sedimentological construction and processes of formation.

According to Hayward (1985), alluvial fans of the Gulf of Aqaba coast can be subdivided into four types: (1) large fans that formed primarily during the Pleistocene, which merged into a braided fluvial system and now are nearly inactive, (2) large Pleistocene fans that have one dominant entrenched channel and accompanying submarine canyon, (3) moderately sized (3 to 4 km long) and active fans, and (4) small wadis that discharge directly into the sea. Along the coastal plains bordering each gulf, alluvial fans of these types are

merging to form thick deposits of poorly sorted debris eroded from adjacent
mountains (Fig. 8). These sediments are dominantly composed of igneous and
metamorphic crystalline basement rock fragments that range in dimension from
large boulders in proximal parts of the fans to gravels, sands, and silts in
the distal parts.

Gvirtzmann and Buchbinder (1978) indicate that the regional slope is
extremely steep (5°-14°, measured from the inland mountain tops to the
submarine deeps) as compared to common worldwide values of 1° to 10°. On the
western coastal plain near the Strait of Tiran, alluvial fans and braided
stream deposits merged from Wadi Kid and Wadi Adawi (Fig. 1) form a broad
depositional surface 3-7 km wide that slopes seaward at 0.5° to 1.0°. Fans
studied by Hayward (1985) near the northwestern limit of the Gulf of Aqaba have
surface slopes that vary 6° to 7° at the apex to 1° to 2° in the lower fan,
with a general downfan decrease in particle size.

Alluvial fans bordering the Aqaba and Suez gulfs build sporadically during
flash floods following infrequent rains. Debris flows from the wadis create
upper-fan areas characterized by large, chaotically distributed blocks (1 m to
3 m in diameter) supported in a matrix of sand- to cobble- sized sediment. In
the middle and lower fan areas, coarse longitudinal gravel bars form in the
active braided pathways of sedimentation like those described by Leopold and
Wolman (1957). These bars extend over most of the fan surface, but they tend to
decrease in relief, longitudinal dimension, and grain size in a downfan
direction. Near the coast, gravel beaches are commonly cemented to form a
coarse-grained beach rock (Fig. 9). Although fine-grained sediments are not
characteristic of these alluvial fan deposits, sandy beaches are common
shoreline features, with eolian dunes migrating inland across coarser deposits
of the lower and middle fan. A detailed description of surficial deposits of
these present-day fans is given by Hayward (1985). In contrast, Neev and Emery
(1967), Nir (1967), and recently Sneh (1979) have described the internal
anatomy of similar fan deltas and accompanying lake deposits now exposed and
dissected along the Dead Sea rift north of the Aqaba Gulf.

In arid and semi-arid alluvial fans, carbonates generally are associated
with the distal fan margins as products of evaporation and chemical
precipitation as well as biogenic activity (Nickel, 1985). This trend is
particularly true of Suez and Aqaba, where sabkhas and coral reefs have
developed at the coast. Perhaps the most interesting features of the
overlapping alluvial coastal plain deposits are the coral reefs that grow on
their seaward ends.

Figure 8. A, Aerial view of alluvial fans building along the northwestern flank
of the Gulf of Aqaba. The small fan is prograding along the southern flank of
the large Pleistocene fan from Wadi Watir at Nuweiba. B, Boulder-and
gravel-sized sediment in the proximal part of an alluvial fan.

112

Fig. 9. Beach rock at the seaward end of an alluvial fan from Wadi Adawi (Gulf of Aqaba). Note the gravel-sized particles comprising this intertidally cemented deposit.

6 CORAL REEFS

Coral reefs of the Red Sea's northern gulfs are primarily of the fringing reef type. Because of considerably deeper water in the Gulf of Aqaba as compared to Suez, offshore slopes are steeper and therefore reefs tend to be directly attached to the coast. Although fringing reefs are also common in the Gulf of Suez, there is a tendency for barrier and patch varieties to develop on the less precipitous transitions from coast to gulf floor. Structural control on these basins and their margins results in the formation of very few broad, shallow areas for reefs to colonize and develop. As a consequence, reefs are generally confined to a narrow belt along shore. Even so, Loya (1972) shows that over 100 scleractinian species from 40 genera are members of reef communities in the Gulf of Aqaba. Guilcher (1955), Friedman (1968), Mergner (1971), Scheer (1971), Erez and Gill (1977), Loya and Slobodkin (1971), Gvirtzmann et al. (1977), and Gvirtzmann and Buchbinder (1978), among others, have described dominant reef types from the northern Red Sea.

6.1 Narrow Shoreline Reef

Figure 10 summarizes the association of fringing reefs with commonly occurring coastal features. In areas where the coastal plain is narrow and terrigenous sediment input to the basin is minimal, reefs with a profile such as shown in Figure 10 (A-A') are typical. These fringing reefs generally have a rather uniform and level reef flat with a steep seaward edge of actively growing corals (Fig. 11A) that drops to depths of approximately 4 m to 10 m before encountering a gentler offshore slope (Fig. 11B). At this point the profile can be quite variable, depending on the regional slope into the basin. However, as illustrated by Mergner and Schumacher (1974), Erez and Gill (1977), Gvirtzmann and Buchbinder (1978), Gabrie and Montaggioni (1982), and others, combinations of reef-derived sediment cover and small-scale coral heads and coral pavements typically comprise the forereef area. Along shore, this general reef-to-basin profile is interrupted by less steep sediment-dominated slopes corresponding to local embayments opposite small wadis. The wadis represent sediment transport pathways from adjacent mountainous terrain. They are seldom active, but their frequency of activity is enough to prevent significant reef growth in the nearshore zone.

Especially along the shallow margins of the Suez Gulf, distal ends of small alluvial fans and braid plain deposits are frequently not colonized by reef-building organisms. Smooth depositional slopes, as shown in Figure 12, occur in shallow-water areas where flash floods have recently buried pre-existing reefs or augmented young fan deposits.

6.2 Reefs Protecting Alluvial Fans

The most diverse reef morphologies tend to form in association with large and relatively old alluvial fans (Fig. 10, B-B'). In this setting (Fig. 13), reefs that have developed along the seaward margins of fans protect the siliciclastic alluvial sediment from being eroded and redistributed by marine processes. As pointed out by Gvirtzmann et al. (1977), these large alluvial systems were initiated during the Pleistocene and are currently being capped by Holocene alluvium and distal fan carbonates. Like fringing reefs along other shoreline types, those forming on alluvial fan deposits have developed a steep seaward face and an offshore profile that commonly incorporates rather large patch reefs.

Because of the diverse topographies of alluvial deposits that functioned as substrates for modern reefs, some reefs take the form of barriers rather than fringing varieties. Smaller scale morphological variations such as deep furrows

114

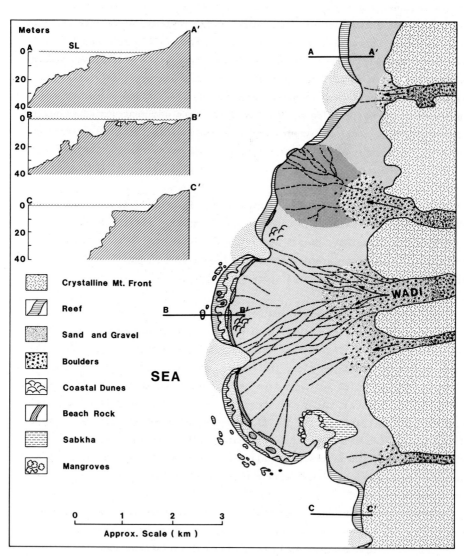

Figure 10. A schematic representation of fringing reefs of the northern Red Sea and their association with commonly occurring coastal features.

Figure 11. A, The actively growing seaward edge of a fringing reef located along the southern Sinai coast of the Aqaba Gulf. Note the nearly vertical profile of the reef and its flat top. B, Sediment-covered slope at the base of steep face of fringing reef. Sediments are composed primarily of in situ carbonate debris at this site.

Fig. 12. Depositional slope at the seaward end of a young alluvial fan building into the southern Suez Gulf near the town of Hurghada (see Fig. 1).

and cavities in the reef flat and seaward reef crest can be followed landward and correlated with alluvial channels of the drainage network etched into the present fan surface (Gvirtzmann and Buchbinder, 1978). The furrows, which apparently originated as erosional cuts made prior to the Holocene transgression, are features that are not to be confused with spur-and-groove topography, also apparent in reefs from both the Gulf of Suez and the Gulf of Aqaba. Sneh and Friedman (1980) show that spur and groove orientations in northern Red Sea reefs are roughly coincident with the paths of refracted waves from the dominant wind direction, a pattern established from other coral reefs throughout the world (e.g. Roberts, 1974).

Where backreef lagoons and broad reef flats have formed isolated pools (occasionally surrounded by mangroves), migrating sand bodies, coral pavements, and patch reefs are common. Like the furrows, the pools, to ~300 m in diameter and ~3 m deep, are related to channels extending from local wadis. Lagoonal shorelines are characterized by sand and gravel beaches, which are commonly lithified into beach rock (Fig. 9). Eolian dunes are present on the coastal parts of the largest fans.

117

Figure 13. A, Alluvial fan at Neviot (Gulf of Aqaba) showing the braided stream
network on the fan's surface and coral reefs protecting its distal end. Note
the large reentrants, which correspond to dominant sediment tansport routes.
The picture was taken looking south along the fault-controlled mountain front
of the Sinai (National Geographic, 1982). B, Wave breaking on a reef that
fringes a large alluvial fan along the south-central Sinai coast of the Aqaba
opposite Wadi Dahab (photo courtesy A. Sneh).

Figures 10 and 13 clearly illustrate that reefs do not completely rim the outer margins of alluvial fans. Reentrants characterized by steep, sediment-floored slopes separate the reefs into well-defined segments. In all cases, non-reef parts of the fan's perimeter correspond to major sediment transport pathways from nearby wadis to the sea. Many of these and other narrow embayments along the Gulf of Aqaba coast are thought to have cut through preexisting fringing reefs and alluvial fans during lower sea level and now represent drowned canyons (Gvirtzmann et al. 1977). Similar features are not so common around the margins of the Suez Gulf because of its shallow basin configuration. When sea level dropped below 70 m, nearly all of the Suez Gulf floor was exposed to subaerial processes, whereas only the basin margins were exposed around the Gulf of Aqaba.

The fact that reef communities have not colonized both large coastal reentrants and relatively small cuts through most fringing and barrier reefs suggests that these sediment transport pathways are still active. Side-scan sonar data from the southern Sinai coasts of the Aqaba Gulf (Roberts, in preparation) confirm that reefs developing on alluvial fan deposits are dissected by well-defined channels of various dimensions (Fig. 15). The steeply inclined channels are pathways for transport of sediment to deep depositional environments. Both in situ produced carbonate sediments and transported siliciclastics are moved through these systems. The largest and best developed channels (Fig. 14) can normally be traced directly onshore to major discharge points of the wadi system. Sediment transport events are frequent enough to prevent coral colonization on coarse debris in the channels. Through time the reef accretes along the channel flanks, producing the appearance of an incised channel. In addition, submersible observations (to >200 m) of canyons opposite wadis of the southern Sinai coast suggest that large volumes of sediment are actively moving down these systems, creating features linked to processes of submarine erosion (H. W. Fricke, 1982, personal communication).

Sediment flux to the coast is sporadic, as dictated by the arid climate. Although large runoff events can transport sufficient sediment to inundate the fan-fringing reefs in terrigenous debris, channels conduct much of the runoff to the forereef. Figure 15 shows turbid water from a flash flood being transported through the reef by way of well-established channels and furrows. Since the proportion of fine-grained suspended load is small, excessively turbid water does not generally persist for long periods of time after a flash flood, thus minimizing the effects of high suspended sediment concentrations on the reef community. In some cases, sand and gravel literally flood the living reef, covering corals of the reef flat and reef crest with a matrix of coarse siliciclastic sediment. Because of the nonuniform sediment delivery to the reef

119

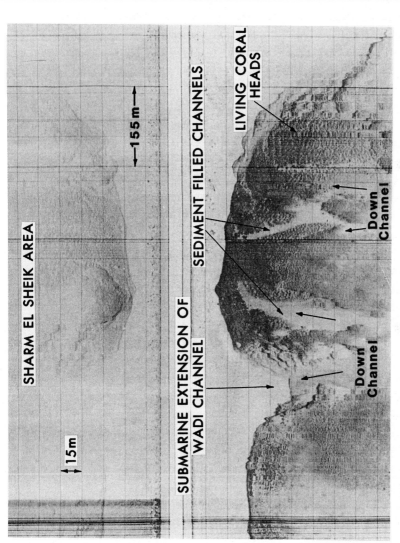

Fig. 14. Side-scan sonograph of a reef that has formed on alluvial deposits opposite a small wadi at Sharm el Sheikh. Note the well-defined sediment transport routes through the reef. The large sediment-floored cut through the reef connects directly to a major channel of the Wadi system. Rough "texture" on the sonograph represents coral colonies of the living reef community.

through the alluvial channel network and varying flood intensity, sediment inundation is an incomplete process that favors rapid recolonization of damaged reef areas. Rapid recovery is probably enhanced by the availability of gravel-sized sediment (>60 mm diameter), which forms an ideal substrate for coral larvae (Hayward, 1982). Gravels transported through channels in the shallow reef to the forereef form deposits that make suitable substrates for patch reef development. In addition to these flash-flood deposits, recent side-scan sonar data (Roberts, in preparation) suggest that slumping of the steep, shallow reef margins, perhaps during periods of lowered sea level, provide large limestone blocks on which modern coral communities colonize.

Pleistocene reefs now exposed as coral terraces along the coast record these alternating episodes of reef growth and siliciclastic sediment inundation (Gvirtzmann and Buchbinder, 1978; Givrtzmann and Friedman, 1977).

6.3 Reefs Along Rocky Coasts

Reefs attached directly to rocky coasts tend to have very steep profiles (Fig. 10, C-C'). In many cases, for example the popular SCUBA diving sites surrounding Ras Muhammad, southern Sinai coast, the shallow reef is actually overhanging, with a near-vertical wall descending into deep water.

Schick (1958) recognized that slopes of the southern Gulf of Aqaba margins, especially submarine slopes south of the Strait of Tiran, are virtual precipices, nearly always with a gradient exceeding 50%. Narrow sediment chutes common to steep-profiled reefs provide avenues of transport to deep sedimentary environments (Fig. 16). The source of sediments for these chutes is primarily in situ produced carbonate debris. Only at sites of embayments in the shoreline where drainage networks emerge at the coast are siliciclastic sediments introduced in significant quantities to the basin. Although slopes are steep, sediments are trapped and stabilized by the low-relief but complicated framework provided by the reef community. Sites of abundant sediment accumulation are breaks in slope that represent submarine equivalents of elevated Pleistocene coral terraces, common to the perimeters of both the Gulf of Suez and the Gulf of Aqaba (Schick, 1958; Givrtzmann and Buchbinder, 1978). Submarine profiles shown in Figure 17 illustrate the terraced configuration of the steep forereef along the rocky southern Sinai coast. As pointed out by Gvirtzmann et al. (1977), during the Wurm glaciation, sea level dropped below the sill depth of approximately 130 m at the Strait of Bab-el-Mandeb, connecting the Red Sea and Indian Ocean. Erosional terraces and nickpoints to this depth can be expected as a consequence of changes in sea level. Not only

Figure 15. Turbid runoff being channelized through the reef at the margin of a large alluvial fan at Neviot (Wadi Watir), along the Sinai coast of the Gulf of Aqaba. Note the sections of reef that appear to be only marginally affected by this event (photo courtesy of A. Sneh).

Figure 16. Narrow sediment chute typical of steep-profile reefs along rocky and narrow coastal plain sectors of the Gulf of Aqaba and Gulf of Suez shorelines.

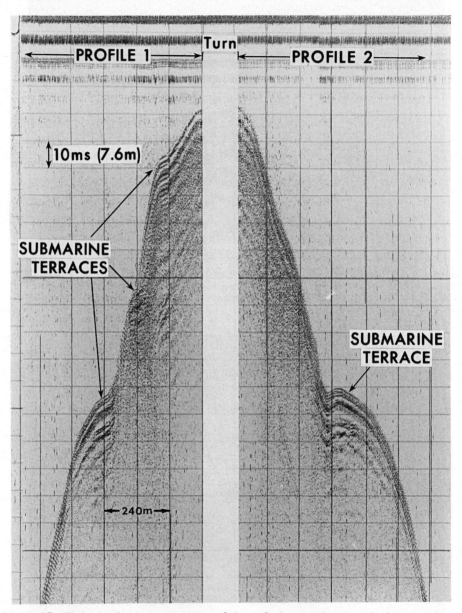

Figure 17. High-resolution seismic profiles of the coral-covered slopes and terraces off the southern Sinai coast near Sharem el Sheikh. Note the distinct submarine terraces, which act as sites for sediment accumulation as well as reef growth. Profiles were run perpendicular to the reef front.

are these breaks in slope sites of sediment accumulation, but high-resolution seismic and echo-sounder profiles suggest they are also preferred sites of reef development.

7 Sabkhas

Coastal sabkhas are not extensive depositional settings of the northern Red Sea area because of basin tectonics, which encourages development of narrow and relatively high sloping coastal plains. Sabkhas do, however, occur in low-sloping coastal embayments, at the ends of flat and inactive alluvial fans, and in interfan areas. In arid coastal environments where sediment input from continental as well as marine sources is low and broad coastal flats are present, evaporation drives precipitation of halite, gypsum, and anhydrite. Offshore sediments are washed over the sabkha during storms, in which the supratidal areas are flooded with marine water.

The broader and lower angle coastal plains bordering the Gulf of Suez are more conducive to development of sabkhas than higher relief coastal plains surrounding the Gulf of Aqaba. However, both gulfs have restricted regions where supratidal sediments include siliciclasti sabkhas with limited evaporite accumulation (Gavish, 1974). Beach rock is a common and nearly continuous feature of the intertidal zone, with carbonate shoals that are oolitic, particularly in the northern Gulf of Suez (Sass et al. 1972) accreting in shallow subtidal areas. Sneh and Friedman (1984) show that spit complexes are common coastal features along most of the Gulf of Suez's Sinai coast. Embayments created by these accretionary features have developed tidal flats and coastal sabkhas. The supratidal sabkhas are located at the distal ends of alluvial fans. Eolian dunes are migrating across the sabkhas in some localities, an association Glennie (1970) described from sabkha-spit complexes of the Persian Gulf.

Zeit Bay (Fig. 18), along the southwestern coast of the Suez Gulf, is a good example of the relationship between coastal embayments and local sabkhas. In this setting the perimeter of the bay is formed chiefly of slowly prograding alluvial deposits derived from local crystalline basement rock highlands. Although reefs locally colonize the shallow alluvial nearshore areas of the bay, the seaward margins of these deposits are frequently deposition slopes similar to the one shown in Figure 12. Small coastal sabkhas have formed behind local beach and berm systems. A large supratidal plain is present at the landward end of the bay. As shown in Figure 19, this well-developed siliciclastic sabkha has the typical algal mat-evaporite association. Zones of lithified dolomitic crusts are common on the upper supratidal surface of this sabkhas as well as others along the southern Suez coast.

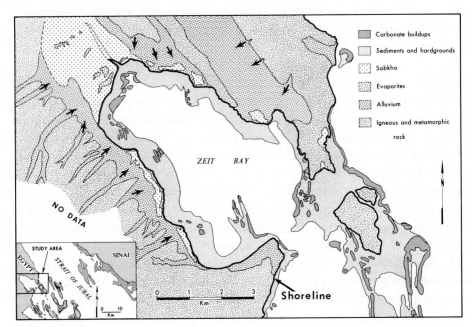

Figure 18. Map of Zeit Bay, along the western coast of the Gulf of Suez. Note the large area of sabkhas at the northern end of the bay. Alluvial sediments eroded from surrounding crystalline rock highlands and small sabkhas form the perimeter of the bay.

Figure 19. A view of the sabkha surface at the northern end of Zeit Bay (see Fig. 18) showing salt-encrusted algal mats. Sediments of the sabkha are primarily terrigenous, containing evidence of recent carbonate cementation and limited evaporite formation.

8 CENTRAL GULF ENVIRONMENTS

8.1 Carbonate Platforms and Islands

As a consequence of Tertiary tectonics that initiated and distinguished the
two gulfs of the northern Red Sea, somewhat different seafloor morphologies and
sediment fills are presently developing. The much shallower Gulf of Suez has
broader coastal plains whose shallow extensions into the basin are the
substrates for a variety of reef types. Mid-gulf shoals, reefs, and hardgrounds
are in direct contrast to the deepwater (>1,000 m) setting for the central Gulf
of Aqaba. Straits that connect these gulfs to the Red Sea are also quite
different. The Strait of Jubal, at the southern end of the Suez Gulf, is wide
(>60 km) and complicated, with numerous reef-ringed islands, shoals, and small
carbonate platforms (Figs. 1 and 6). In contrast, the Gulf of Aqaba's Strait of
Tiran is narrow (~5 km) and constricted by an elongate reef complex
approximately 3 km long, which appears to be developing on a fault-controlled
block (Figs. 1 and 4).

Of particular interest are the numerous northwest-trending islands, reefs,
and carbonate platforms in the southwestern Strait of Jubal. Even though the
basic submarine topography of the Gulf of Suez is subdued, southern flanks of
the Jubal Strait are complicated by many topographic irregularities, most of
which are capped or ringed by reefs. Garfunkel and Bartov (1977) show that the
shoals off the southernmost Sinai coast are crossed and probably controlled by
numerous faults trending in a northwest direction. Shoal complexes of the
central Gulf, sites of the large Morgan and Belayim offshore oil fields, have
been confirmed by drilling as reflecting subsurface structural highs (Glen
Steen, GUPCO, 1981, personal communication). Despite rapid sedimentation,
continuing tectonic activity has a profound effect on the modern sea floor of
the Suez Gulf. Although high-relief islands such as Shadwan (Fig. 1), with
exposed crystalline basement rocks as well as Neogene sediments, are probably
fault controlled, many smaller islands, carbonate platfforms, and shoals along
the western Strait of Jubal are not easily linked to faults or other subsurface
controls. Garfunkel and Bartov (1977) indicate that numerous gulf-parallel
normal faults cut the Pliocene-Recent alluvium flanking the present gulf.
Perhaps offsets on the sea floor by this recent tectonism functioned as a
control for the initiation of reef growth and later modifications of submarine
topography. Roberts and Murray (1984) suggest that tidally dominated flow in
conjunction with sediment transport has also imposed important modifications on
seafloor morphology.

The Ashrafi reef complex (Fig. 6) is typical of various carbonate buildups
that are developing on the shallow western flank of the southern Suez Gulf and
is similar in some morphological characteristics to reefs located in the Strait
of Tiran. Figure 20 illustrates that both large carbonate platforms and small

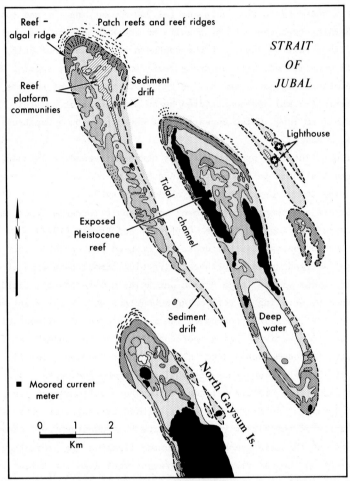

Fig. 20. A map view of the developing carbonate platforms and shoals in the
Ashrafi reef complex showing the general distribution of reefs, platform top
communities (primarily brown algae), sediment, and Pleistocene remnants (mapped
from air photographs, with major subdivisions confirmed by field
observations).

shoals have streamlined shapes and a northwest alignment that is consistent with the Suez Gulf's axis. The flow field surrounding these developing platforms was monitored by in situ recording current meters moored in Ashrafi channel and the adjacent Strait of Bughaz. Results from these meters (Fig. 21) indicate a strong tidal fluctuation, with the Bughaz record clearly showing a

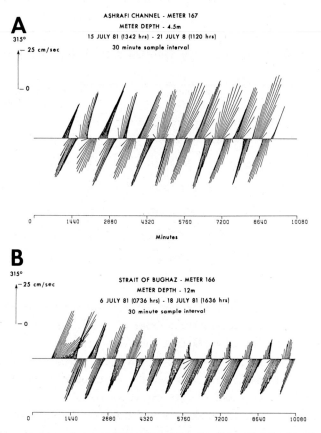

Fig. 21. Time series current vectors. A, Ashrafi Channel. B, Strait of Bughaz. Note the strong tidal periodicity in both records and slightly stronger flow to the south in the Bughaz record (a function of augmentation by southerly directed wind stress). Northerly flow is above horizontal axis.

net flow to the south, resulting from augmentation by southerly directed wind stress. Unfiltered data display typical current speeds of >50 cm/sec. The southerly directed winds and waves, in conjunction with strong tidal exchange, produce net sediment transport to the south. This dynamical setting promotes a wave-dominated northern end to the platforms and islands, with a downdrift or southerly end characterized by sediment accumulation.

Because of steady northerly winds (8-15 m/sec), northern ends of the Ashrafi platforms are characterized by reef communities that have adjusted to relatively high levels of wave energy. Even though the Gulf is a shallow, fetch-limited basin, coincidence of the dominant wind direction with the gulf axis produces an effective maximum fetch of 250 km. Resulting waves are focused on the northern ends of the islands, carbonate platforms, and reefs in the southern Gulf of Suez area. Basic procedures described by Bretschneider and Reid (1954) and Bretschneider (1966) were used to construct Table 1, which

TABLE 1. COMPARISON OF CARIBBEAN AND SOUTHERN GULF OF SUEZ SHORELINE WAVE ENERGY LEVELS

Site	Geography	Wave Energy Levels (ergs/m/sec)	
Nicaraguan	Western Caribbean	4.3×10^{10}	(South)
	(Fetch 2000 km)	1.2×10^{10}	(North)
Grand Cayman	Central Caribbean	4.2×10^{10}	(Windward)
	(Fetch 700 km)	0.3×10	(Leeward)
Ashrafi Reef	Southern Gulf Suez	4.6×10^{8}	(Avg from N)
	(Fetch 250 km)	3.4×10^{9}	(20% from N)

shows that wave energy calculations for the northern end of Ashrafi Reef average 4.6×10^{8} ergs/m/sec, just one order of magnitude less than wave power calculated for mid-Caribbean reefs that receive the full force of trade wind seas (Roberts, 1974). During about 20% of the year, maximum winds produce waves in the southern gulf that create wave energy levels on the Ashrafi reefs comparable to those calculated for mid-Caribbean reefs.

Windward (northern) platform margins are characterized by rough bottom topography (relief generally less than 3 m) representing the growth of reef masses separated by localized areas of sediment accumulation and hardgrounds (Fig. 23). From the wave-stressed areas to the lower energy platform flanks, reef masses diminish systematically in size and abundance. Roberts and Murray (1984) indicate that, even though levels of wave energy in the Jubal Strait approach that of the central Caribbean (Table 1), well-organized spur-and-groove topography is not present in the reef-dominated parts of the Ashrafi platforms.

Unlike fringing reefs of the northern Red Sea, these platforms do not have a
well-developed reef flat with a steep reef front on the seaward side. The
Ashrafi platforms are characterized by a coral-covered reef crest that slopes
gently into the forereef, compared to steep fringing reefs of the basin
margins. In the wave-stressed areas, the bottom is typically covered with a
thick growth of Cymodocea, an important part of the bottom community to a depth
of 15-20 m, where Halophila, another noncalcareous marine plant, becomes
dominant.

Abundant coral growth is characteristic of the windward margins of platforms
and islands. Corals that commonly inhabit these environments are Acropora,
Porites, Platygyra, Pocillopora, Alcyonium, and Stylophora. Coralline algae are
abundant as both encrusting and free-living forms.

Narrow tidal channels separate platforms, shoals, and islands of the
southern Suez Gulf. The straits of Bughaz and Jubal are the boundaries that set
the Ashrafi complex apart from other islands and shoals (Fig. 20). Platform
margins adjacent to these channels are sites of hardgrounds and areas of patchy
reef growth separated by well-defined routes for off-platform sediment
transport (Fig. 22). Flanks of these elongate platforms have a thick sediment
cover. Bioturbation by the burrowing shrimp Callianassa is abundant to the
point of excluding benthic plant communities. In areas where the burrowing and
mound-building process is not so intense, the marine plant Halophila is
abundant. Roberts et al. (1981) found that mound building by Callianassa can be
important in the sediment transport process, even in the presence of low-speed
currents. Sediment ejected into the water column is advected downdrift by the
mean flow. Favored southerly flow in the southwestern Suez Gulf (Fig. 21)
suggests that mean sediment transport will be north to south, a process that
appears responsible for a long tail of sediment on the westernmost Ashrafi
platform (Fig. 20).

Platform tops in the Ashrafi complex are shallow, flat, and wide (1.5-2.0
km) at the wave-stressed northern ends. To the south, the western platform
narrows and deepens. Sediment cover becomes thicker, and hardgrounds plus reef
patches are common. The shallow parts of the platform are characterized by a
benthic community of mainly noncalcareous plants and animals. Sargassum and
some noncalcareous green algae are common inhabitants of the platform top. Very
few corals and only infrequent occurrences of calcareous green algae (Halimeda
in the platform edge reef community) were encountered. A possible explanation
for this conspicuous absence of calcareous greens and corals may be related to
high salinities, 47 ‰ on the platform top. Water of this elevated salinity
contrasts sharply with the 40-41 ‰ water of the gulf.

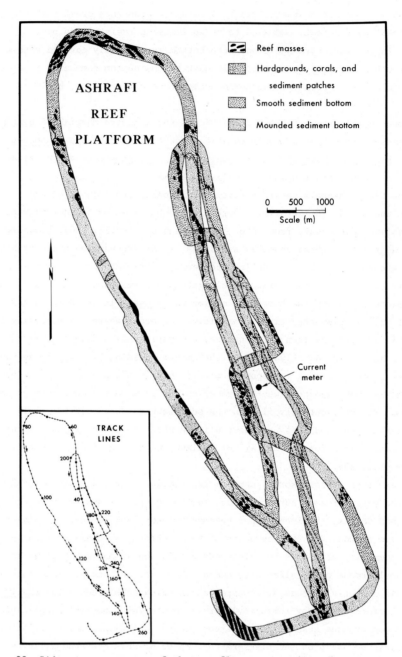

ASHRAFI REEF PLATFORM

Reef masses

Hardgrounds, corals, and sediment patches

Smooth sediment bottom

Mounded sediment bottom

0   500   1000
Scale (m)

Current meter

TRACK LINES

Figure 22. Side-scan sonar map of the sea floor surrounding the westernmost carbonate platform in the Ashrafi reef complex (Fig. 20). The inset shows the survey track lines and navigation fix points.

9 Sediments

Figure 23 summarizes constituent particle data from 16 bottom samples
collected on an east-west transect across the Strait of Jubal north of the
Ashrafi reefs. It is interesting to note that results of these data show that
modern sand- and silt-sized sediments of the Gulf floor are mostly carbonates.
Shallow flanking environments of the strait contain up to 7% noncarbonate
components, mostly quartz in the fine to coarse silt range. In the deep strait,
noncarbonates account for less than 4% of the silt- to sand-sized components.
Emery (1963) found the same general conditions in the Gulf of Aqaba, where he
describes the chief component as calcium carbonate. Even in this much deeper
gulf, modern sediments are primarily carbonate muds with discrete silts and
sands of both carbonate and siliciclastic origin, interpreted as turbidites
(Emery, 1963). These calcareous marine shales and coarser turbidites are common
source rock/reservoir associations in many ancient petroleum-rich basins.

Along the shallow western flank of the Suez Gulf (Fig. 24, samples 1-3),
foraminifera tests (mainly Peneroplis and Amphistegina) comprise up to 75% of
the coarse fraction (Fig. 24A). However, in the deep Jubal Strait, foram tests
account for only 25%, while intraclasts (commonly including microfauna tests)

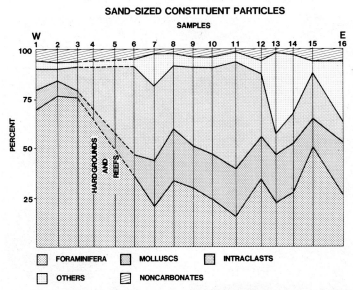

Fig. 23. Major constituent particles in the sand fraction of 16 sediment
samples collected on a transect across the Jubal Strait ~10 km north of the
Ashrafi reef complex. Data derived from point counts (200 points/thin
section).

132

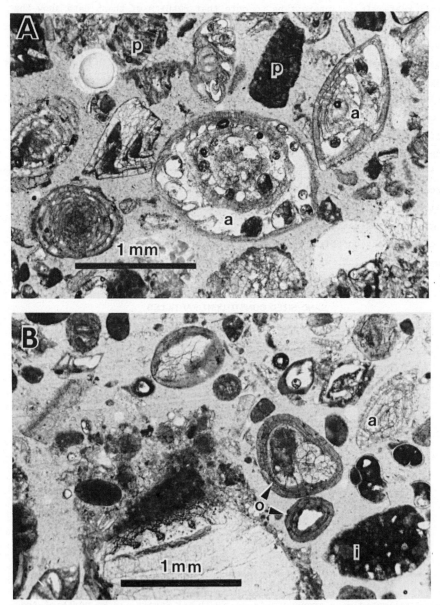

Figure 24. Thin-section photomicrographs of samples from the Gulf of Suez. A, Sediments of the western Gulf (sample 2, Fig. 23) are rich in foraminifera. _Amphistegina_ (a) and _Peneroplis_ (p) are the most abundant. B, Central Gulf sediments (sample 10, Fig. 23) contain less foraminifer and more composite grains, oolites (o) and interclasts (i).

and mollusc fragments account for over 50% of the silt- and sand-sized components. Locally, coral and coralline algae grains are important, especially at the topographic break between the shoulder of the Gulf and the deep strait. Oolites are present in both shallow and deep sediments (Fig. 24B), but they are not abundant. Oolites appear to be forming on the shallow nearshore shelf (Sass et al. 1972; Sneh and Friedman, 1984) and are probably transported to deep water by eolian processes.

Sediments of the platform top are composed of three main constituents, coral fragments, coralline algae, and foraminifera tests. Minor constituents consist of Halimeda fragments, mollusc debris, and various calcareous spicules (mainly from echinoderms and alcyonarians). A preliminary survey of sediments from the platform flank, much of which is transported off the top, indicates 130 species and subspecies of Foraminifera and 42 species of ostracoda (Earl Manning, LSU, 1985, personal communication). The microfauna is quite distinctive because of the relatively high frequency of occurrence of large forams such as Peneroplis and Amphistegina and thick-walled porcelaneous varieties such as Borelis. This trend is typical of highly saline environments where the process of calcification is facilitated. Most of the foraminifera are benthic varieties. The planktonics are sparse and are mostly juvenile forms. Platform flank sediments as described above from samples collected at depths of approximately 30 m differ from those of the shallow platform top. The deep samples are more diverse, with a few planktonic species. Elphidium, Peneroplis, Amphistegina, Sorites, and miliolids are most common on the shallow platform. The reduction in species on the platform top is perhaps a response to elevated salinities.

10 SUMMARY

The geologically young tectonic troughs of the northern Red Sea, Gulf of Aqaba and Gulf of Suez, are sites of active siliciclastic and carbonate sediment deposition. Tertiary crustal movements set the stage for basin definition and the filling process. Even though the two basins have a somewhat different structural heritage and present bathymetric configuration, depositional processes and facies relationships are very similar. Structural fabric of the Suez Gulf cuts through stable margins of the Arabo-Nubian Shield. Northwest-southwest trending normal faults form the graben, which has subsequently filled with sediment to a maximum thickness of over 5 km. In contrast, the Gulf of Aqaba represents a southern extension of the Dead Sea rift, which is considered to be a leaky transform fault. Perimeters of both gulfs consist primarily of tilted and elevated blocks of crystalline basement rock. Erosion of these flanking mountains has produced sediment for the basin-filling process as well as construction of narrow coastal plains composed primarily of overlapping alluvial fans.

Although tectonic controls have produced the geologic framework for sedimentation, climate and physical processes have played important roles in determining characteristics of the sedimentary fill as well as facies relationships. Figure 25 summarizes some of the important impacts of an arid climate on basins of deposition such as the Gulf of Suez and Gulf of Aqaba. Lack of abundant rainfall (10-70 mm/yr) leads to unvegetated sediment source areas, like the mountainous Sinai Peninsula. Because soils lack humic components and plant cover to absorb moisture from infrequent rains, runoff is maximized and often results in flash floods. These sporadic and sometimes intense events are primary mechanisms of sediment transport from source area to the basin. Resulting sedimentary accumulations of boulders, gravel, and sand are the numerous alluvial fans that are prograding from the mountain fronts directly into the marine environment.

Low atmospheric moisture in arid climates promotes intense evaporation and high radiative heat loss, resulting in extreme temperature gradients. Evaporation rates in the northern Red Sea gulfs are among the world's highest.

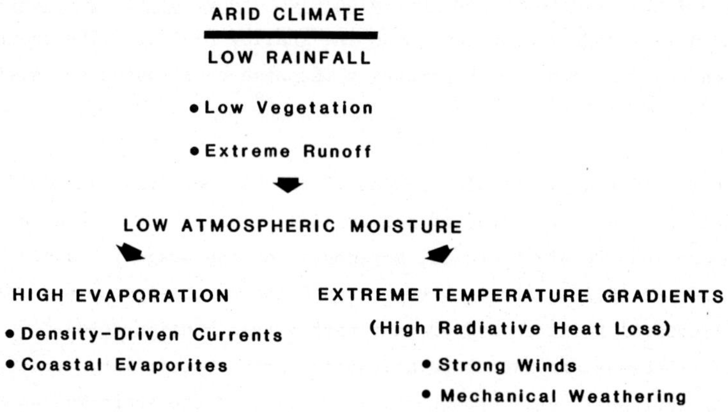

ARID CLIMATE

LOW RAINFALL

- **Low Vegetation**

- **Extreme Runoff**

LOW ATMOSPHERIC MOISTURE

HIGH EVAPORATION

- **Density-Driven Currents**
- **Coastal Evaporites**

EXTREME TEMPERATURE GRADIENTS
(High Radiative Heat Loss)

- **Strong Winds**
- **Mechanical Weathering**

Fig. 25. Flow diagram of the important influences of an arid climate on sedimentary systems like those of the northern Red Sea.

Sea-surface water loss results in increased salinities and compensatory surface flow from the larger Red Sea to replace water lost to the atmosphere. Strong density-driven currents capable of sand-sized sediment transport are generated by the evaporative process. Thick accumulations of evaporites can be deposited in these basins when straits connecting them to larger water bodies become so restricted from tectonic movements and/or reef growth that evaporation far exceeds replenishment with normal sea water. Additional effects of high evaporation are to enhance the formation of coastal sabkhas, where abundant evaporite minerals are deposited.

Heating and cooling of mountain fronts bordering the gulfs help to create a strong mountain-valley wind that essentially blows down the axis of each elongate water body. Along with regional winds, which are steered by the topography, local winds resulting from daily temperature gradients create significant surface wind drift currents and energetic waves. Coincidence of the dominant wind direction with the long axis of each gulf results in utilization of the maximum fetch in generating waves. Consequently, downwind ends of these basins experience high levels of wave energy.

Extreme temperature gradients are also important to siliciclastic sediment production in arid climates. Processes of mechanical weathering of rocks are efficiently accomplished under these conditions.

Relationships between siliciclastic and carbonate facies in arid tectonic basins, the gulfs of Aqaba and Suez, are summarized in a block diagram (Fig. 26). Basins of this description are characterized by rapid changes in both topography and sedimentary facies from the marginal fault scarps to the central basin. Transitions from siliciclastics to carbonates occur over distances of a few kilometers. In the case of reefs along the Aqaba and Suez gulf margins, transitions from coastal plain alluvium to reef can occur over a distance of <1 km.

Since major pulses of alluvial fan sedimentation are infrequent, reefs actively colonize distal ends of the fans, which interface with the marine environment. Studies of the northern Red Sea area (Friedman, 1982; Roberts and Murray, 1984; Hayward, 1985; and others) have shown that basin perimeter reefs protect the distal ends of alluvial fans from being eroded by marine processes and maintain steep basin margins. Channels through the reefs, ranging from major embayments (described as inherited Pleistocene features by Gvirtzmann et al. 1977) to well-developed furrows, provide access to the basin for silici-clastic sediments during normal runoff events. This process of sediment transport effectively bypasses the reefs, with only minimal damage to the reef community. Gravel carried to the reef flat and forereef by flash floods

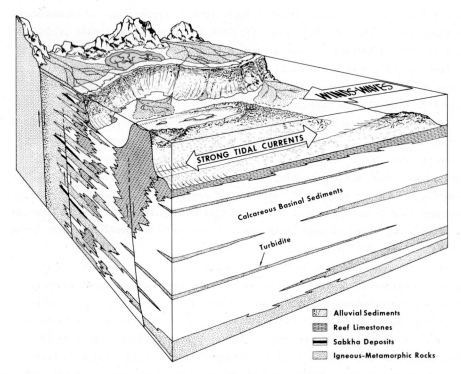

Fig. 26. Block diagram summarizing the structural and facies relationships typical of arid tectonic basins such as the Gulf of Aqaba and Gulf of Suez.

provides substrates for rapid recolonization of the damaged coral community, a process that can also be documented in ancient reefs from arid settings (Gvirtzmann and Buchbinder, 1978; Hayward, 1982). Siliciclastics from small flash floods, plus the contribution of carbonate sediments produced by the reef community, keep the transport pathways through the reef open and relatively free of coral growth. Submersible observations (Drs. H. Fricke and J. Erez, 1982, personal communication) and side-scan sonar surveys suggest that gravity-driven creep processes keep sediment in these pathways constantly moving downslope, thus eliminating substrates for sedimentary organisms. The disproportionate quantities of fine-grained terrigenous sediment available for transport to the basin suggests that damage to coastal and other nearshore reefs by inundation with terrigenous muds is minimal in arid settings. Therefore, fan deltas and coral reefs are compatible and are frequently associated with depositional systems.

Periods of increased tectonic activity result in a dominance of siliciclastic input to the basin, which may result in burial of reefs and widespread occurrence of terrigenous sands distributed by basinal processes. Gulf-parallel sand bodies in the Miocene fill of the Gulf of Suez seem to have been distributed by strong bidirectional currents in the elongate basin (Stan Slocki, GUPCO, 1981, personal communication). Periods of relative tectonic inactivity favor development of substantial carbonate structures, abundant carbonate sediments, and sabkha evaporites. Both the siliciclastic and carbonate sediments can be good reservoirs, while sabkha evaporites are among the best seals. Dutton (1982) describes a Pennsylvanian age sequence of rocks from the subsurface of the Texas Panhandle that consists of alternating limestones and siliciclastic sandstones similar to the nearshore facies of the Aqaba and Suez gulfs. The Texas rocks represent alluvial fans and algal mound limestones. Both the limestones and siliciclastic sandstones are productive reservoirs when in a favorable structural position.

Central parts of tectonic basins, especially like the Gulf of Suez, can contain favorable reservoir, source, and seal relationships. As Figure 26 shows, carbonate platforms may form in conjunction with dominantly fine grained calcareous basinal sediments and relatively thin siliciclastic as well as carbonate turbidites. Abdine and Kanes (1983) indicate that the Gulf of Suez is Egypt's most important hydrocarbon-producing area. Most of the production is from the Cenozoic, which consists of alternating siliciclastics and carbonates with a thick evaporite unit (Fig. 2). Faulting has produced a variety of structural and structural-stratigraphic traps. As previously discussed, subsurface structures are reflected on the present Gulf floor by shoals and reefs.

In summary, studies of modern processes of sedimentation and their resultant depositional features in the tectonic gulfs of Aqaba and Suez have led to the following conclusions about siliciclastic-carbonate facies relationships in these warm, arid settings:

1. Basin structure and variations in tectonic activity control relief on basin flanks, siliciclastic sediment discharge, and alternations between carbonate and siliciclastic dominance in the basin fill.

2. An arid climate promotes sporadic but intense transport events of dominantly coarse siliciclastic debris, resulting in progradation of alluvial fans from fault-controlled mountain fronts directly into the marine environment.

3. Infrequency of runoff to the basin allows carbonates, especially coral reefs, to develop in shoal areas and on the distal ends of alluvial fans. Reefs protect alluvial deposits from being eroded by marine processes, plus they maintain steep slopes into the basin with well-defined through-reef sediment transport routes. Because of the steep fault-controlled basin margins, most reefs are of the fringing variety.

4. Abrupt lateral as well as vertical facies changes are to be expected in these arid tectonic basins. Transitions from alluvial fan to coral reef can occur over distances of <1 km.

5. Reefs and abundant siliciclastic sedimentation are compatible in an arid setting for the following reasons:

   a. Large pulses of siliciclastic sediment are infrequent, allowing interim periods for carbonates to develop.

   b. Siliciclastic sediments are primarily coarse gravels to sands, with only a small silt-clay fraction. Detrimental effects to reef organisms by fine-grained sediment is minimal. Gravels function as ideal substrates for colonization by coral larvae. Even in areas where siliciclastics have overwashed the reef, recolonization and stabilization of the sediments are rapid processes. Gravel deposits on the forereef can be the substrates for patch reefs.

   c. Reefs contain well-defined transport routes, large channels and furrows, that connect landward with alluvial streams so that siliciclastic sediments are efficiently carried through the reef to the basin.

6. Basin sediments under present conditions are primarily foraminifer-rich carbonate muds with both siliciclastic and carbonate turbidites/debris flows.

7. Straits that connect these tectonic gulfs with a larger water body such as the Red Sea can be areas where tidal currents and waves are focused. Reef growth coupled with sediment transport in this setting can lead to initiation of shoals, reefs, and the development of carbonate platforms. Continued accretion of these features leads to further restriction of the strait, increasing salinities in the gulf, and perhaps eventual deposition of evaporites.

8. Sediments that can function as source rocks, reservoirs, and seals
are all a part of the fill sequence in these tectonic basins.
Numerous unconformities and complex structures provide most
hydrocarbon traps, as evidenced by production in the Gulf of Suez.

ACKNOWLEDGEMENT

Financial support for these studies was provided by the Coastal Sciences
Program of the Office of Naval Research. Drs. A. R. Bayoumi (Director of
Institute of Oceanography, Cairo) and G. F. Soliman (Acting Director of
Physical Oceanography, Alexandria) helped coordinate the project in the Gulf of
Suez. The authors would also like to acknowledge personnel at Gulf of Suez
Petroleum Company (GUPCO) for their helpful discussions about geology of the
Suez Gulf. The U.S. Navy Oceanography Office, Israeli Navy (Cmdr. J.
Troestler), and the Israel Oceanographic and Limnological Research Institute
(Dr. Artur Hecht) provided invaluable logistical and technical support for the
Strait of Tiran study. Technical assistance from CSI was provided by Rodney
Fredericks, Norwood Rector, and Walker Winans. Crew members of the R. V.
Shikmona, Haifa, are gratefully acknowledged for their skills and professional
execution of the project. The Geological Survey of Israel is very gratefully
acknowledged for providing critical pieces of geophysical equipment for the
project. Celia Harrod and Gerry Dunn are acknowledged for drafting
illustrations for this paper. Kerry Lyle provided the photographic work.

REFERENCES CITED
Abinde, S., and W. Kanes, 1983. Egyptian exploration, background, and future
    potential. Oil and Gas Journal, 71-72.
Assaf, G., and J. Kessler, 1976. Climate and energy exchange in the Gulf of
    Aqaba. Monthly Weather Review, 104, 381-385.
Ben-Avraham, Z., 1984. Structural framework of the Gulf of Elat (Aqaba) -
    northern Red Sea (abs.). EOS Transactions, American Geophysical Union, 65,
    423.
_____, G. Almagar, and Z. Garfunkel, 1979. Sediments and structure of
    the Gulf of Elat (Aqaba) - northern Red Sea. Sedimentary Geology, 23,
    239-267.
Bretschneider, C.L., 1966. Wave generation by wind, deep and shallow water,
    In: A.T. Ippen (Editor), Estuary and coastal hydrodynamics, New York,
    McGraw-Hill, pp. 133-196.
_____, and R.O. Reid, 1954. Modification of wave height due to bottom
    friction, percolation, and refraction. U.S. Army Corps of Engineers Beach
    Erosion Board Technical Memorandum 45, 1-36.
Dutton, S.P., 1982. Pennsylvanian fan-delta and carbonate deposition,
    Mobeetie Field, Texas Panhandle. AAPG Bulletin, 65, 389-407.
Emery, K.O., 1963. Sediments of the Gulf of Aqaba (Eilat), In: R.L. Miller
    (Editor), Papers in marine geology, Shepard Commemorative Volume, New York,
    MacMillan, pp. 257-273.

Erez, J., and D. Gill, 1977. Multivariate analysis of biogenic constituents in Recent sediments off Ras Burka, Gulf of Elat, Red Sea. Mathematics and Geology, 9, 77-98.

Friedman, G.M., 1968. Geology and geochemistry of reefs, carbonate sediments, and waters, Gulf of Aqaba (Elat), Red Sea. Journal of Sedimentary Petrology, 38, 895-919.

_____, 1982. Coexisting terrigenous sea-marginal fans and reefs at the shores of the Gulf of Aqaba (abs.). Hamilton, Ontario, 11th International Congress on Sedimentology, 109.

_____, and J.E. Sanders, 1978. Principles of sedimentology. New York, Wiley, pp. 792.

Gabrie, C., and L. Montaggioni, 1982. Sedimentary facies from the modern coral reefs, Jordan Gulf of Aqaba, Red Sea Coral Reefs, 1, 115-124.

Garfunkel, Z., 1981. Internal structure of the Dead Sea leaky transform (rift) in relation to plate kinematics. Technophysics, 80, 81-108.

_____, and Y. Bartov, 1977. The tectonics of the Suez Rift. Bulletin No. 71, Geological Survey of Israel.

Gavish, E., 1974. Geochemistry and mineralogy of a Recent sabkha along the coast of Sinai, Gulf of Suez. Sedimentology, 21, 397-414.

Gilboa, Y., and A. Cohen, 1979. Oil trap patterns in th Gulf of Suez. Israel Journal of Earth Sciences, 28, 13-26.

Glennie, K.W., 1970. Desert sedimentary environments. Developments in Sedimentology No. 15, Amsterdam, Elsevier, 222.

Griffiths, J.F., and K.H. Soliman, 1972. The northern desert. In: World survey of climatology, v. 10 (Climates of Africa). New York, Elsevier, pp. 75-131.

Guilcher, A., 1955. Geomorphologie de l'estremite septentrionale du banc corallien Farsan, in Resultats scientifiques de campagnes de la Calypso, 1, Campagne 1951-1952 en Mer Rouge. Annales de Institut Oceanographique, 30, 204.

Gvirtzmann, G., and G.M. Friedman, 1977. Sequence of progressive diagenesis in coral reefs. Studies in Geology 4, AAPG, 357-380.

_____, and B. Buchbinder, 1978. Recent and Pleistocene coral reefs and coastal sediments of the Gulf of Eilat. In: Post Congress Guidebook, 10th International Sedimentological Congress, 163-189 p.

_____, _____, A. Sneh, Y. Nir, and B. Friedman, 1977. Morphology of Red Sea fringing reefs: a result of the erosional pattern of the last glacial low stand sea level and the following Holocene recolonization. Memoirs du Bureau de Recherches Geologiques et Minieres, 89, 480-491.

Hayward, A.B., 1982. Coral reefs in a clastic sedimentary environment: fossil (Miocene, S.W. Turkey) and modern (Recent, Red Sea) analogues. Coral Reefs, 1, 109-114.

_____, 1985. Coastal alluvial fans (fan deltas) of the Gulf of Aqaba (Gulf of Eilat) Red Sea. Sedimentary Geology, 43, 241-260.

Kanes, W.H., and S. Abdine, 1983. Egyptian exploration: backgrounds, models, and future potential. Book of Abstract, American Association of Petroleum Geologists Annual Convention, Dallas, 104 pp.

Leopold, L.B., and M.G. Wolman, 1957. River channel patterns: straight, meandering, and braided. U.S. Geological Survey Professional Paper 282-B, 39-85 pp.

Loya, Y., 1972. Community structure and species diversity of hermatypic corals at Eilat, Red Sea. Marine Biology, 13, 100-123.

_____, and L.B. Slobodkin, 1971. The coral reefs of Eilat (Gulf of Eilat, Red Sea). In: D. R. Stoddart and M. Yonge, (Editors), Regional variation in Indian Ocean coral reefs, Symposium of the Zoological Society of London. London, Academic Press, 28, 117-139.

Mart, Y., and J.K. Hall, 1984. Structural trends in the northern Red Sea. Journal of Geophysical Research, 89, 11, 352-11, 364.

Mergner, H., 1971. Structure, ecology, and zonation of Red Sea reefs. In: D. R. Stoddart and M. Yonge (Editors), Regional variation in Indian Ocean coral reefs, Symposia of the Zoological Society of London. London, Academic Press, 28, 141-161.
_____, and H. Schumacher, 1974. Morphologie, Ukologie, and Zonierung von Korallenriffen bei Aqaba (Golf von Aqaba, Roten Meer). Helgolander Wiss. Meeresunters, 26, 238-358.
Morcos, S.A., 1970. Physical and chemical oceanography of the Red Sea. Oceanographic and Marine Biology Annual Reviews, London, George Allen and Unwin, 8, 73-202.
Murray, S.P., and A.L. Babcock, 1982. Observations of two-layered circulation in the Gulf of Suez (abs.). San Francisco, American Geophysical Union Annual Meeting, December 1982.
_____, A. Hecht, and A. Babcock, 1984. On the mean flow in the Tiran Strait in winter. Journal of Marine Research, 42, 265-287.
National Geographic, 1982, 161, 4, 450-451.
Neev, D., and K.O. Emery, 1967. The Dead Sea, depositional processes and environments of evaporites. Geological Survey of Israel Bulletin, 147 pp.
Neumann, J., 1952. Evaporation from the Red Sea. Israel Exploration Journal, 2, 153-162.
Nickel, E., 1985. Carbonate in alluvial fan systems, an approach to physiography, sedimentology, and diagenesis. Sedimentary Geology, 45, 83-104.
Nir, Y., 1967. Some observations on the morphology of the Dead Sea wadis. Israel Journal of Earth Sciences, 16, 97-103.
Purser, B.H., 1982. Carbonate-silicate transitions in the Arabo-Persian Gulf (abs.). Hamilton, Ontario, Abstracts of the IAS Meeting, 110 pp.
Roberts, H.H., 1974. Variability of reefs with regard to changes in wave power around an island. Brisbane, Australia, 2nd International Coral Reef Symposium, 2, 497-512.
_____, in preparation, Origins of patch reefs on the Tiran sill (northern Red Sea) as suggested by side-scan sonar data. Geology.
_____, and S.P. Murray, 1983. Gulfs of northern Red Sea: depositional settings of distinct siliciclastic-carbonate interfaces (abs. + oral presentation). Book of Abstracts, Annual AAPG Convention, Dallas, Texas, 541 pp.
_____, and _____, 1984. Developing carbonate platforms: southern Gulf of Suez, northern Red Sea. Marine Geology, 59, 165-185.
_____, Wm. J. Wiseman, Jr., and T.H. Suchanek, 1981. Lagoon sediment transport: the significant effect of Callianassa bioturbation. Manila, 4th International Coral Reef Symposium, 1, 459-465.
Sass, E., Y. Weiler, and A. Katz, 1972. Recent sedimentation and oolite formation in the Ras Matarma Lagoon, Gulf of Suez, In: D. J. Stanley, (Editor), The Mediterranean Sea. Stroudsburg, Pa., Dowden, Hutchinson and Ross, pp. 279-292.
Scheer, G., 1971. Coral reefs and coral genera in the Red Sea and Indian Ocean. In: D. R. Stoddart and M. Yonge (Editors), Regional variation in Indian Ocean coral reefs: London, Academic Press, pp. 329-367.
Schick, A.P., 1958. Tiran: the straits, the island, and its terraces. Israel Exploration Journal, 8, 120-130, 189-196.
Shinn, E.A., 1973. Sedimentary accretion along the leeward, SE coast of Qatar Peninsula, Persian Gulf. In: B. H. Purser (Editor), The Persian Gulf. New York, Springer-Verlag, pp. 199-209.
Sneh, A., 1979. Late Pleistocene fan deltas along the Dead Sea rift. Journal of Sedimentary Petrology, 49, 541-552.
_____, and G.M. Friedman, 1980. Spur and groove patterns on the reefs on the northern gulf of the Red Sea. Journal of Sedimentary Petrology, 50, 981-986.

142

_____, and _____, 1984. Spit complexes along the eastern coast of the Gulf of Suez. Sedimentary Geology, 39, 211-226.

U.S. Naval Oceanographic Office, 1965. Sailing directions for the Red Sea and Gulf of Aden. Publication 61, 375.

Vercelli, F. 1931. Nuove recherche sulli correnti marine del Mar Roso. Genova, Annali Idrografici, 12, 1-74.

Chapter 5

TERRIGENOUS AND CARBONATE SEDIMENTATION IN THE GREAT BARRIER REEF PROVINCE

A.P. BELPERIO and D.E. SEARLE Geological Survey of South Australia, P.O. Box
151, Eastwood, S.A. 5063, (Australia); Geological Survey of Queensland, P.O.
Box 194, Brisbane, Qld. 4001 (Australia)

ABSTRACT
The mixed carbonate-terrigenous shelf system associated with the Great Barrier
Reef covers an area of about 270,000 km^2. Fluvial sediment delivery to the
Province averages 10 tonnes per metre of coastline per year and dominates
modern sedimentation on the shelf. The northward dispersion of sediment from
individual river mouths results in a merging of coastal interdeltaic deposits
and an inner shelf zone dominated by terrigenous mud sedimentation.
Intermittent high turbidity on the inner shelf precludes significant reef
construction. Modern fluvially derived sand is largely contained in the
littoral zone, and the terrigenous sand and gravel component of middle- and
outer- shelf sediments are of relict or palimpsest origin. Coastal
sedimentation accounts for the bulk of the Holocene terrigenous sediment
budget, with vertical accumulation at rates of up to 8.5 mm/yr and seaward
progradation at rates of up to 4 mm/yr. Holocene reef growth has occurred at
rates as high as 16 mm/yr, and the reef complexes represent a significant
reservoir of autochthonous carbonate sediment. However, their contribution of
allochthonous carbonate to the shelf is limited. In situ skeletal carbonate
production on the middle- and outer- shelf floor is locally significant;
however, seismic profiles and sea-bed cores confirm that much of the shelf
floor is devoid of major Holocene sediment accumulation. Quaternary sea-level
fluctuations have promoted terrigenous-carbonate intermixing, but have
suppressed long-term carbonate production. The changing locus of maximum
terrigenous sedimentation (the coastal zone) associated with fluctuating sea
levels has resulted in an overwhelming dominance of terrigenous clastics in the
subsurface of the shelf.

1 INTRODUCTION

The Great Barrier Reef Province, adjacent to over 2000 km of coastline,

forms one of the world's largest carbonate-terrigenous shelf systems. The area

of continental shelf occupied by the province is approximately 270,000 km^2, of

which about 80,000 km^2 (30%) of the province is dominated by modern terrigenous

sedimentation (50% CaCO3), with the remainder of the shelf floor either

sediment free or mantled by a veneer of mixed carbonate and reworked sediment.

Investigators working in the province have long recognized a meridional

zonation in regional shelf sediment lithofacies (Maxwell, 1968; Maxwell and

Swinchatt, 1970; Marshall, 1977; Marshall and Davies, 1978; Orme and Flood,

1980). The patterns of sedimentation broadly reflect a western (mainland)

terrigenous source, an eastern (reef) carbonate source, and a central lagoon

floor that shows only minor influence from either side. Although the patterns

of surficial sediment distribution have been recognized for some time, most

researchers have concentrated their investigations on the reef carbonate tract,

144

and hence have failed to appreciate the significance of terrigenous
sedimentation in the context of shelf evolution. The objectives of this paper
are to document the contribution of biogenic, bioclastic and terrigenous
clastic components of Holocene shelf sedimentation in the Great Barrier Reef
Province, to provide a framework for assessing Quaternary shelf evolution.

2 GEOLOGICAL BACKGROUND

The Queensland continental margin is a relatively young geological feature
that evolved from a rifting and sea floor spreading phase during the Paleocene,
in the process creating the Coral Sea Basin (Taylor and Falvey, 1977; Mutter
and Karner, 1980; Falvey and Mutter, 1981). Margin subsidence and
progradational shelf construction continued throughout the Tertiary, with reef
growth commencing probably no earlier than the Pleistocene (Symonds et al.,
1983).

Block faulting and elevation of the continental interior accompanied
subsidence of the shelf (DeKeyser, 1964; DeKeyser et al., 1965; Paine, 1972;
Ollier, 1978). Concomitant erosion and scarp retreat during the Tertiary and
Quaternary led to the accumulation of a sequence of erosion products that make
up the present day coastal plain. This coastal plan, of alluvial and colluvial
origin, is well developed along the north Queensland coastline (Fig. 1), and is
continuous with the shallow continental shelf (regional gradient 0.0008).

Quaternary sea-level fluctuations have resulted in periodic shelf
emergence and submergence and in outer shelf reef growth. The Holocene
transgression interrupted alluvial processes on the shelf between 11,000 and
7000 years BP, and present sea level had essentially stabilized by 6000 years
BP (Belperio, 1979a; Chappell, 1983; Hopley, 1983, Thom and Roy, 1983). Initial
outer-shelf reef growth lagged behind the transgression, but accelerated as the
turbid coastal zone shifted progressively landward. With stabilization of sea
level, many reefs built up rapidly to low water to become net sediment
contributors to the surrounding shelf floor. Along the coast, depositional
progradation over the last 6000 years has resulted in extensive marginal marine
sedimentation, forming a quasi-continuous low-gradient (0.0002) coastal marine
plain, in places more than 10 km wide.

The present-day reef province can be classified as a rimmed continental
shelf in which a 2000 km-long tract of shelf-edge reefs restricts water
circulation and wave action within the adjacent shelf lagoon. There are significant latitudinal variations in shelf morphology and bathymetry, and shelf
width varies from 300 km in the south to 24 km in the north near Cape Melville
(Maxwell, 1968). The terms "inner," "middle," and "outer" shelf are used to

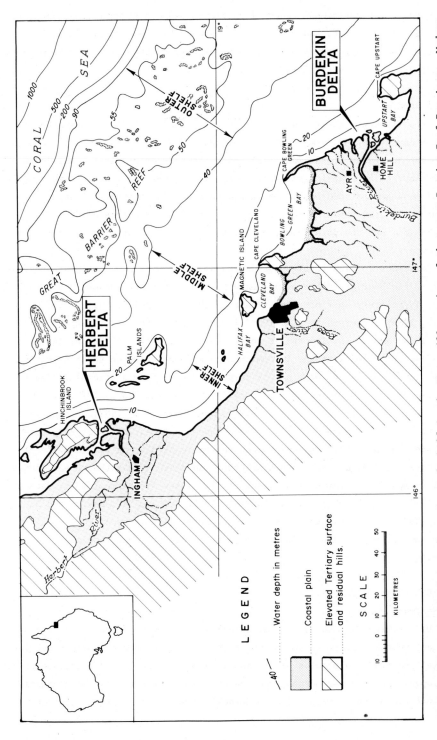

Fig. 1. The coastal plain and continental shelf in the Townsville region of the great Barrier Reef Province. Modern terrigenous sedimentation extends to the 20 m isobath. From Belperio (1983a).

denote the three meridional zones of surficial sedimentation (terrigenous, mixed carbonate/relict, and reefal). In the central Great Barrier Reef Province, these zones correspond to water depths of approximately 0-20 m, 20-40 m, and 40-80 m, respectively (Fig. 1).

3 DATA BASE

Establishing a general shelf sedimentation model requires knowledge of present day types and rates of sediment input, operative dispersal processes, and modes and rates of sediment deposition. It further requires a knowledge of their temporal and spatial variability and of the resultant three-dimensional lithofacies generation. Despite an enormous expansion in the past decade, marine geological research in the Great Barrier Reef Province needs integration, and the data required to establish a sedimentation model remain fragmentary.

Sedimentation and terrigenous facies development along the inner shelf and coast of the Barrier Reef Province have been studied in Edgecumbe Bay (Frankel, 1971), in Princess Charlotte Bay (Frankel, 1971; Chappell et al., 1983; Chivas, et al., 1983; Torgersen et al., 1983) in Broad Sound (Burgis, 1974; Cook and Mayo, 1978) and in the Townsville region (Belperio, 1978, 1979a, b, 1983a, b) (See Fig. 2 for localities). Information on sedimentation on the middle- and outer-shelf floor is scanty, and is largely deduced from surficial sediment samples and from continuous seismic profiles (Orme and Flood, 1977; Orme et al., 1978a, b; Searle et al., 1980, 1981; Johnson et al., 1982; Searle, 1983a, b; Symonds, 1983; Symonds et al., 1983; Johnson and Searle, in press). More data are available for the reef carbonate complexes dealing with aspects of bioclastic production, sediment transport, and growth rates (Smith and Kinsey, 1976; Davies, 1977; Davies and Kinsey, 1977; Kinsey and Davies, 1979; Harvey et al., 1979; Kinsey, 1981; Davies and West, 1981; Frith, 1983; Davies and Hopley, 1983). However, these data are specific to particular reefs or reef environments, and extrapolation to the entire reef province remains conjectural.

4 TERRIGENOUS SEDIMENTATION

4.1 Sediment sources

Direct fluvial input is by far the greatest source of modern terrigenous sediment to the reef province. Except locally, other sediment sources such as coastal and headland erosion and eolian input are minimal. Estuaries are poorly developed at most river mouths, and the numerous rivers act as point sources of sediment to the shelf (Fig. 2). The Burdekin River is the largest present-day

147

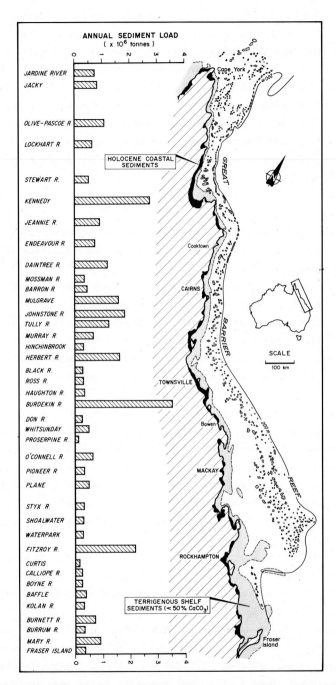

Fig. 2 The average annual sediment discharge of river systems draining into the Great Barrier Reef lagoon. Areas of Holocene coastal progradation and terrigenous shelf sedimentation are indicated. Specific localities referred to in the text are also shown. From Belperio (1983b).

contributor of terrigenous clastics to the reef province with an average annual
sediment load of 3.0×10^6 tonnes of mud and 0.45×10^6 tonnes of sand
(Belperio, 1979b). The 28×10^6 tonnes of sediment contributed annually by all
rivers draining into the reef province is equivalent to a delivery rate of 10
tonnes per metre of coastline per year (Belperio, 1983b).

Relict terrigenous sediment is also incorporated into modern, actively
accumulating shelf floor sediment, particularly where the mantle of Holocene
sediment is thin and patchy. The contribution is difficult to quantify,
although the relict component is easily recognised by the intense pitting,
weathering and ferruginization of grains.

4.2 Sediment transport

The only study dealing specifically with terrigenous sediment transport is
that by Belperio (1978). Tide, wind and river discharge are identified as the
primary controls on sediment dispersion.

Rainfall along the North Queensland coast is strongly seasonal, and river
runoff occurs in a series of short-lived floods, mainly during the months
December to March. River plumes are advected northwestwards along the coast by
the cross-shelf density gradient, and are contained within the inner shelf zone
(Belperio, 1978; Wolanski and Jones, 1981). Most sediment brought by floods is
deposited close to river mouths and is subsequently redistributed by coastal
currents.

The tide range at Springs varies from 2.5 to 9 m along the North
Queensland coastline, and results in significant intertidal exposure of a
variety of depositional environments (Fig. 3). Tidal currents are weak and
variable over much of the shelf, but increase substantially in shallow coastal
embayments, where they generate a semi-diurnal cycle of erosion and entrainment
of bottom sediment (Fig. 4). Residual transport of the suspended sediment
landwards results in sedimentation in, and progradation of, intertidal coastal
environments (Belperio, 1978).

Persistent southeast trade winds from March to October maintain a well
mixed water column (Pickard, 1977; Walker, 1981b) and generate a coherent
northwestward (alongshore) current in the lagoon of the Great Barrier Reef
Province. Associated wind waves of up to 2.5 m height (Walker, 1981a) suspend
large concentrations of fine sediment on the inner shelf (Fig. 5). The turbid
water is confined near the coast, generally within the 25 m isobath, and a
large northwestward sediment flux results (Belperio, 1978). Variability in the
strength of the southeast trade winds allows reversals in the mean drift of
shelf water to occur, but the low turbidities associated with these "calms"
preclude significant sediment transport.

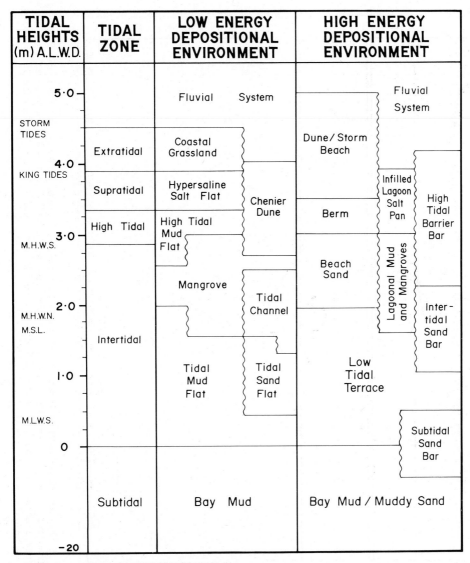

TIDAL HEIGHTS (m) A.L.W.D.	TIDAL ZONE	LOW ENERGY DEPOSITIONAL ENVIRONMENT	HIGH ENERGY DEPOSITIONAL ENVIRONMENT
5·0		Fluvial System	Fluvial System
STORM TIDES — Extratidal	Extratidal	Coastal Grassland	Dune / Storm Beach
4·0 — KING TIDES	Supratidal	Hypersaline Salt Flat	Chenier Dune	Berm	Infilled Lagoon Salt Pan	High Tidal Barrier Bar
3·0 — M.H.W.S.	High Tidal	High Tidal Mud Flat	Beach Sand
2·0 — M.H.W.N. M.S.L.	Intertidal	Mangrove	Tidal Channel	Lagoonal Mud and Mangroves	Inter-tidal Sand Bar
1·0		Tidal Mud Flat	Tidal Sand Flat	Low Tidal Terrace
M.L.W.S. 0			Subtidal Sand Bar
	Subtidal	Bay Mud	Bay Mud / Muddy Sand
−20			

A.L.W.D. .... Above Low Water Datum
M.H.W.S. .... Mean High Water Springs (2·9 m)
M.H.W.N. .... Mean High Water Neaps (1·95 m)
M.S.L. .... Mean Sea Level (1·59 m)
M.L.W.S. .... Mean Low Water Springs (0·4 m)

Fig. 3 The relationship between tide level, wave energy, and depositional environments for the Townsville coastal zone. From Belperio (1978, 1983a).

150

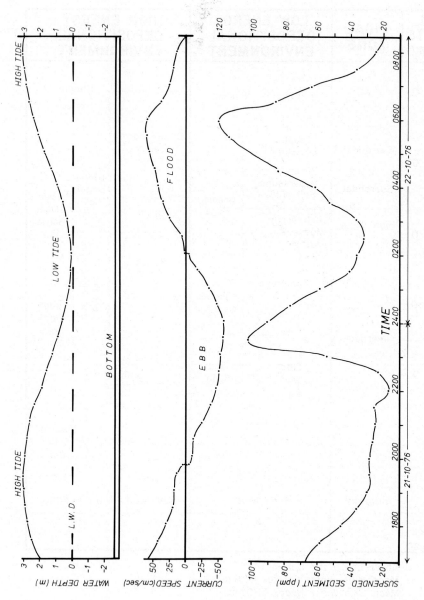

Fig. 4. An example of the semi-diurnal entrainment of suspended sediment by tidal currents at a shallow water site in Bowling Green Bay, south of Townsville. The depth- averaged net flux of particulate matter at this locality was 1.1 tonnes landward per metre of water column for the 12-hour tidal cycle. For further information on monitoring techniques, see Belperio et al. (1983).

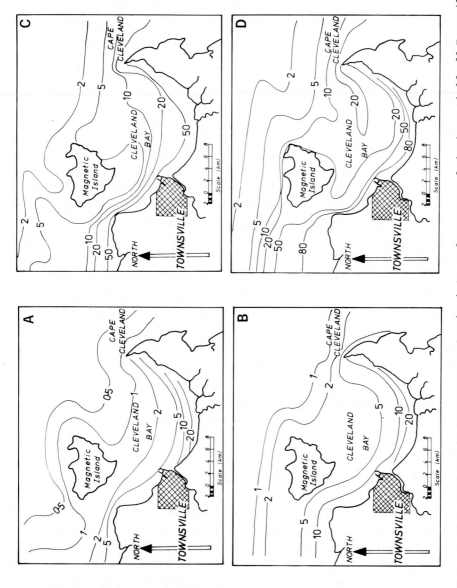

Fig. 5. Contours of suspended sediment concentration (ppm) in surface water of the inner shelf off Townsville during different sea conditions: A) sea smooth; B) sea slight; C) sea moderate; D) sea rough.

Cyclones produce multi-directional, high-energy pulses on the shelf, in contrast to the largely unidirectional trade winds, but they are infrequent and have little long term effect on sedimentation. Minimal dispersion of terrigenous sediment occurs across the shelf. This is reflected in abrupt facies changes, at between 20-25 m water depth and usually less than 20 km offshore, from mud dominated terrigenous clastics to carbonate or palimpsest sediment. The long term resultant of predominately coastwise and coastward sediment transport has been the development of extensive coastal interdeltaic deposits updrift of river mouths (Fig. 6).

4.3 Sediment accumulation

Terrigenous sedimentation dominates the inner continental shelf and coast of the Great Barrier Reef Province. A subordinate contribution of skeletal carbonate detritus, commonly less than 30% by weight, is derived from the in situ biota (Fig. 7A and B). Abrupt changes in sediment facies occur between littoral, inner shelf and middle shelf depositional regimes. Near Townsville for example, sandy littoral-zone sediments merge seaward with mud or sandy mud facies of the inner shelf (Fig. 8A and B). At between 20 and 25 m water depth, the carbonate content increases to over 50%, the terrigenous mud content decreases abruptly, and a significant component of gravelly sand (relict) is introduced. This transition to a relict sand facies marks the limit of signif-icant terrigenous mud accumulation and corresponds to the maximum depth of bottom sediment resuspension by waves which are normally generated by the southeast trade winds (Belperio, 1978). It also corresponds to a change in benthic biota, from an infaunal community adapted to a soft and shifting bottom to an epifaunal community composed predominantly of suspension feeders (Arnold, 1980). Encrusting calcareous algae becomes abundant seaward of the 25 m isobath (Birtles and Arnold, 1983), and the carbonate component of bottom sediment increases progressively to values of over 90% in inter-reef areas of the outer shelf. The terrigenous component of shelf sediments in these areas is largely of relict origin.

Sediment cores on the shelf off Townsville indicate that the Holocene marine deposits are thin and patchy (Davies et al., 1983). The terrigenous mud facies of the inner shelf is also a veneer, largely less than 1 m thick, but sediment thickness increases coastward, where it exceeds 8 m in the littoral zone (Belperio, 1978; 1983a). Site-specific accumulation rates on the shelf are mostly less than 0.4 mm/yr, whereas accumulation rates are an order of magnitude higher in intertidal environments (Fig. 9). Belperio (1978) concluded that intertidal deposition updrift of the Burdekin River mouth accounts for 80% of the terrigenous sediment budget on the shelf near Townsville. Rapid

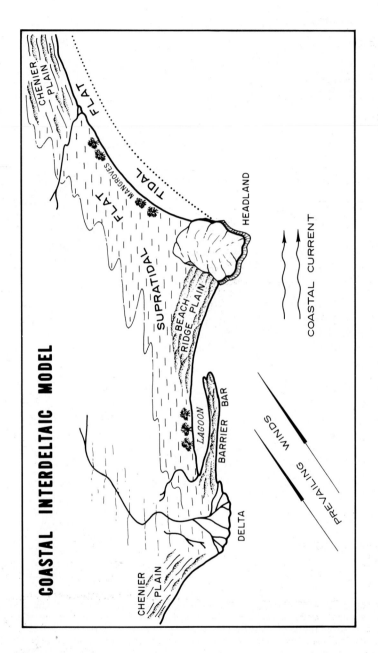

Fig. 6. The major interdeltaic depositional environments of North Queensland. These environments occur in a variable association updrift (northwestward) of most river mouths. From Belperio (1983b).

154

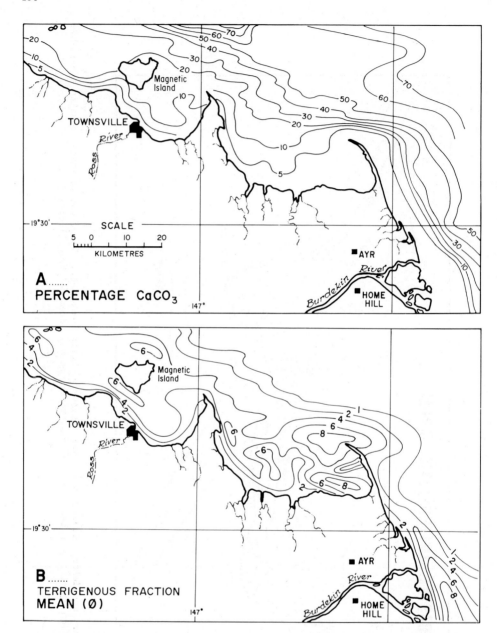

Fig. 7 A) Contours of acid soluble carbonate in bottom sediments of the inner
and middle shelf off Townsville. The 30% carbonate contour approximately
follows the 20-m isobath, while the 50% contour approximates the 25-m
isobath. B) The mean grain size (phi units) of the terrigenous component
of bottom sediments. The inner shelf is largely dominated by terrigenous
mud (4 phi), while the middle shelf is mantled with a mixture of
terrigenous gravelly sand (relict) and bioclastic debris. From Belperio
(1978, 1983a).

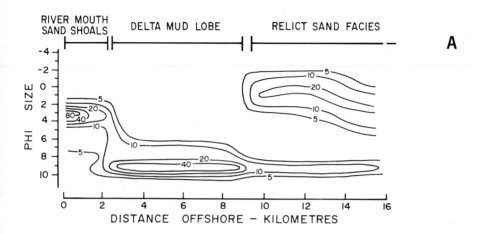

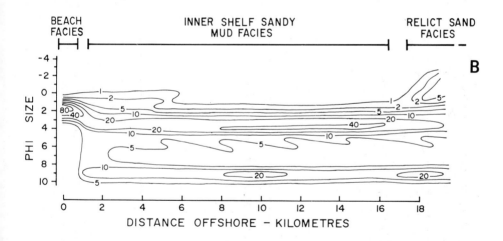

Fig. 8. Grain-size spectra maps of the terrigenous component of bottom
sediments (Dowling, 1977) for shore-normal transects off: A) the Burdekin
River mouth; and B) Halifax Bay, north of Townsville. In both cases, note
the abrupt changes between sandy littoral deposits, mud-dominated facies
of the inner shelf, and the middle shelf, relict gravelly sand facies.
Note the subordinate population of fine clay in the relict gravelly sand
facies which results from minor across-shelf diffusion of suspended
terrigenous mud. From Belperio (1978, 1983a).

156

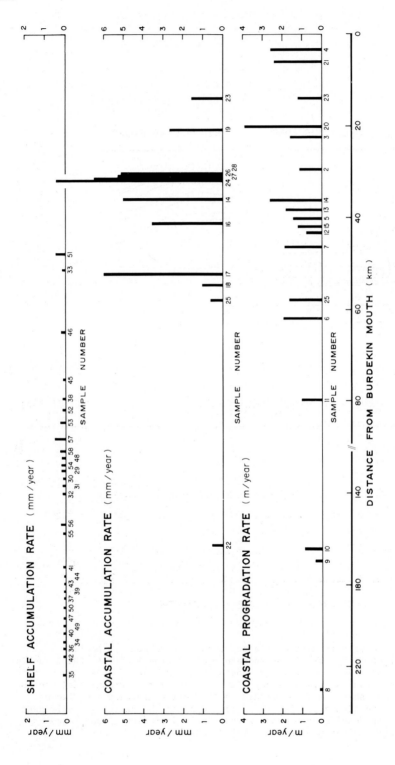

Fig. 9. Accumulation and progradation rate data for the inner shelf and coast in the Townsville–Burdekin region of the Great Barrier Reef Province. Note the limited accumulation on the inner shelf compared with accumulation rates in intertidal environments. Note also the decrease in coastal progradation rate with distance (northwestward) away from the Burdekin River mouth. For data and sample localities, see Belperio (1983b).

intertidal deposition generates a regressive coastal sediment wedge with a low
gradient surface planated to local supratidal level. The principal styles of
coastal progradation and resulting strata development near Townsville are shown
in Figure 10.

Coastal progradation has been greatest in Princess Charlotte Bay, where
over 20 km of outbuilding has occurred in the past 6000 to 7000 yrs and the
resulting coastal sediment wedge exceeds 9 m in thickness. Within the bay, the
terrigenous mud facies extends to the 20 m isobath (Frankel, 1974) and is
generally less than 4 m thick (Searle and Hegarty, 1982; Orme, 1983). Modern
accumulation rates of up to 6 mm/yr have been recorded in this area (Torgersen
et al., 1983; Chivas et al., 1983).

There are numerous other reports which document the significance of
terrigenous coastal marine sedimentation. In the Lizard Island area, the
coastal sand wedge is about 10 m thick and merges seaward into a 2 m thick
layer of mud and sand (Orme, 1983). In the Cairns area, coastal sediments are
up to 20 m thick (Searle, 1979), and terrigenous sands are confined to a narrow
coastal belt extending less than one kilometre seaward (Jones and Stephens,
1983). The terrigenous mud facies extends to 11 km offshore, to water depths of
20-29 m (Maxwell and Swinchatt, 1970) and has a maximum thickness of 7 m
(Searle et al., 1980). In the southern Great Barrier Reef, the coastal sediment
wedge is up to 18 m thick in Edgecumbe Bay (Searle et al., 1981) and 5-10 m
thick in Broad Sound (Burgis, 1974; Cook and Mayo, 1978). The shelf is very
wide in this region, and Maxwell and Swinchatt (1970) considered much of the
sediment mantle on the shelf to be of relict origin, derived in part from
ancient dune systems and older fluviatile deposits.

The predominance of terrigenous sedimentation along the coast and inner
shelf of the Great Barrier Reef Province results from the interaction between
sediment sources and the dynamic conditions on the shelf. Fluvially-derived
sediment is dispersed updrift of river mouths and contributes to progradation
of coastal interdeltaic complexes. The coalescing of such coastal complexes has
resulted in a quasi-continuous coastal marine plain along much of the North
Queensland coastline, that parallels a meridional pattern of terrigenous
deposition on the adjacent shelf (Fig. 2). The limited across-shelf dispersion
of sediment results in abrupt facies changes at between 20-25 m water depth.

5 CARBONATE SEDIMENTATION

Reef carbonate complexes, comprising platforms, reefal lagoons, patches,
shoals and talus deposits, form the most conspicuous part of the carbonate
dominated sector of the shelf. Within the main zone of reef growth, there are
some 2140 individual coral reefs with an overall planimetric surface area of

158

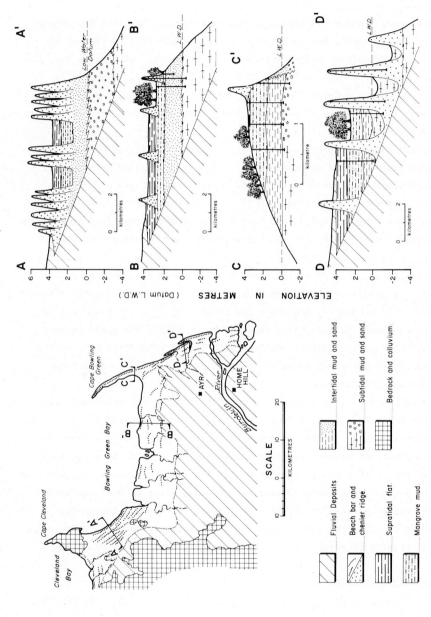

Fig. 10. Styles of coastal progradation and representative cross-sections along the Townsville-Burdekin coastal plain. A) beach-ridge plain; B) chenier plain; C) mangrove forest; D) longshore bar-lagoon complex. For additional information on coastal environments, see Belperio (1978, 1983a) and Pringle (1983).

about 19,000 km^2, representing 7% of the total shelf area (Hopley et al., unpublished data). The postglacial carbonate budget for the shelf can be conveniently divided into autochthonous reef growth, attrition and redistribution of reefal material onto the shelf, and in situ skeletal carbonate accumulation on the shelf floor. All of these sources contribute to shelf buildup, but an undetermined amount escapes the shelf edge and contributes to continental-slope deposition. Relict carbonate may also be locally important where it is exposed on the sea floor.

Postglacial reef growth was initiated between 8000 and 9000 years BP throughout most of the reef province (Davis and Hopley, 1983), and Holocene reef thickness generally varies between 4 and 25 m (Harvey, 1980; Harvey and Hopley, 1981). In the central province, reef foundations are deeper, and the average Holocene reef thickness may exceed 20 m (Davies and Hopley, 1983). A first-order estimate from thickness and area data suggests that the reef complexes represent autochthonous accumulation, averaged over the past 8500 years, of about 30 x 10^6 tonnes/yr. Site-specific data on reef growth rates are more difficult to extrapolate with confidence because of the enormous temporal and spatial variability evident in the measurements. Davies and Hopley (1983) report rates of 1-16 mm/yr, with modes of around 7-8 mm/yr and 4-6 mm/yr for framework components of patch reefs and windward margins respectively. No systematic latitudinal trends are evident in the range of growth rates. Most reefs, however, record a decrease in growth rate close to sea level, which may either reflect a real decrease in carbonate production or result from increasing erosion. Many reefs in the Great Barrier Reef Province have been found to follow a similar growth scheme. Growth is initially vertical and, once sea level has been reached, the reef progrades to leeward (Marshall and Davies, 1982). It is during this lateral progradational phase, particularly once lagoon infill has been completed, that bioclastic detritus may be shed onto the shelf floor in significant amounts.

Biodetrital budget studies at One Tree Reef in the south of the province have included measurements of calcification rates in producing areas and sedimentation rates in sediment-receiving areas (Davies, 1977; Kinsey and Davies, 1979; Davies and West, 1981). Sedimentation in sediment-receiving areas (lagoon, leeside edge, and reef-flat sand sheets) at an average rate of 1300 g/m^2/yr implies a total bioclastic production by One Tree Reef of about 8500 tonnes/yr. Calcification rate measurements show a degree of modality and latitudinal uniformity, and may be rate-limited throughout much of the reef province (Smith and Kinsey, 1976; Kinsey and Davies, 1979; Kinsey, 1981). If so, it may be possible to extrapolate the bioclastic flux data to other reefs. One Tree is a small reef complex covering an area of 9 km^2 and has a main platform that reached sea level about 2000 years ago. Extrapolating the sedimentation rate figures to the entire reef province implies a bioclastic production rate by reefs of about 14×10^6 tonnes/yr. As much of the biodetritus on reef flats contributes to framework growth, the actual contribution of allochthonous detritus to the shelf by reefs must be significantly less than this. Davies and Symonds (1983) estimate the total sediment shed from reefs is of the order of 0.3 tonne per metre of reef edge per year, nearly 60% of which is organic carbon. These figures imply a total contribution by reefs of allochthonous carbonate to the shelf of about 2.5×10^6 tonnes/yr.

Continuous seismic profiles confirm the existence of reef talus deposits extending onto the shelf floor (e.g. Searle, 1983a). However, these deposits are largely restricted to the immediate vicinity of reefs, and there is little apparent dispersal away from the sediment wedges. Indeed, Flood et al., (1978) suggest 2 km as a limit for reefal influence on shelf sediments.

Data on rates of in situ carbonate productivity and bioclastic accumulation on the shelf floor are also sparse. Total carbonate content of sediment is insignificant (<10%) in intertidal and nearshore areas and is low (<50%) on the inner shelf (Fig. 7A). On the middle- and outer-shelf floor, the carbonate component usually comprises 50 to 90% by weight of the bottom sediment. Carbonate productivity on the shelf floor is expected to be at least an order of magnitude less than for the adjacent reef ecosystems, in the range of 10^2 to 10^3 g/m^2/yr (cf. reef calcification rates of 10^3 to 10^4 g/m^2/yr). However, the limited accumulation on middle- and outer-shelf floors, evident from seismic records (Figs. 11 and 12; Searle et al., 1981; Davies, 1983), suggests even lower values. In situ skeletal carbonate production at rates of about 10^2 g/m^2/yr would contribute about 0.1 mm/yr to vertical accumulation of the shelf floor, equivalent to an annual input of about 15×10^6 tonnes to the shelf.

There are some areas of the shelf where much greater productivity and accumulation of bioclastic debris occur. In the northern Great Barrier Reef in particular, luxuriant Halimeda meadows have resulted in biohermal accumulation on the shelf floor in the lee of ribbon reefs (Orme et al., 1978b; Orme and Flood, 1980; Flood and Orme, 1983). Productivity measurements are unavailable for Halimeda vegetation other than measurements by Drew (1983) within reef lagoons off Townsville. Rates of bioclastic production were estimated at about 2230 g/m^2/yr. Such rates are approaching the productivity of entire coral reef systems, hence the accumulation of discrete Halimeda banks up to 20 m thick on the outer shelf north of Cooktown (Orme, 1983). Such sediment build-up is

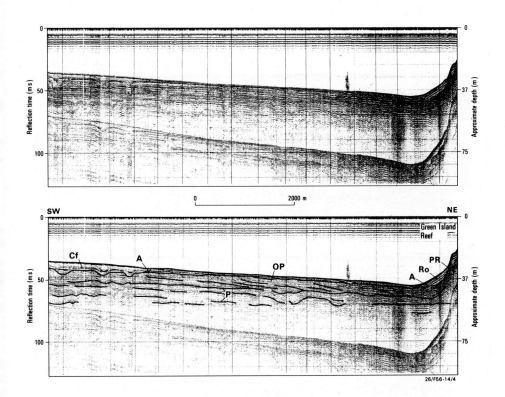

Fig. 11 Seismic reflection profile, and interpretation, from the middle to outer shelf off Cairns (16°45'S). The pre-Holocene surface (A) is fluvially incised and overlain by a terrigenous mud wedge (Cf) which passes seaward into a thin veneer of mixed carbonate and relict quartz sand. Near Green Island Reef, surface A is overlain by Holocene reef and reef talus (Ro). Surface A is underlain by Pleistocene reef rock (PR) and parallel oblique prograded delta-front deposits (OP). Several incised erosional unconformities (P) have resulted from subaerial exposure during late Quaternary, low sea-level phases. This section clearly illustrates the dominance of terrigenous over carbonate deposition, both surficially and in the subsurface of the shelf. From Searle (1983a).

162

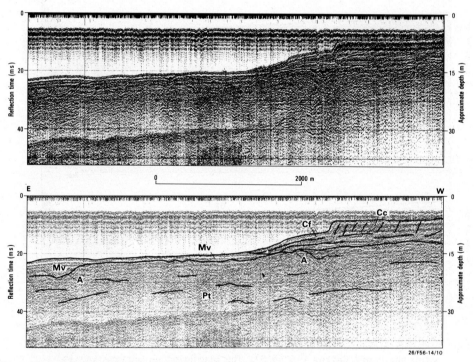

Fig. 12 Seismic reflection profile, and interpretation, from the inner shelf
off the Herbert River delta (18°34'S). During the postglacial
transgression, sediment was reworked and spread across the shelf as a
thin, patchy veneer (Mv) overlying a pre-Holocene surface of exposure (A),
which has resulted from subaerial exposure of terrigneous sediments (Pt).
Since sea level reached its present height, fluvially derived sediment has
built out from the coast as delta front (Cc) and prodelta (Cf) deposits.
From Searle (1983a).

localized, however, and the available seismic data indicate that little or no accumulation has occurred over much of the middle- and outer-shelf floor of the Great Barrier Reef Province during the Holocene high stand of sea level.

6 SEISMIC EVIDENCE

A number of seismic surveys, principally high-resolution boomer, have been undertaken in the Great Barrier Reef Province and provide additional information on sediment thickness and subsurface lithofacies of the shelf. A shallow and prominent subbottom reflector of regional extent has been recognized on high-resolution, continuous seismic reflection profiles (CSP) throughout the reef province. This reflector, commonly referred to as reflector A, is believed to represent the surface of subaerial exposure during the last glacial sea level low (Johnson et al., 1982). Seismic stratigraphic analyses (Searle and Hegarty, 1982; Johnson and Searle, in press) and preliminary results of both published (Orme, 1983; Davies et al., 1983) and unpublished (Geological Survey of Queensland) vibracoring programs on the shelf indicate that the deposits overlying surface A are postglacial in age (late Pleistocene and Holocene).

6.1 Postglacial sequences

Sediments overlying reflector A include fluvial channel-fill deposits of the last glacial low stand, estuarine fill and other patchy transgressive deposits, and various lithofacies of the high sea level stand of the past 6000 years.

6.1.1 Low sea level and transgressive deposits

The oldest postglacial seismic facies in the Great Barrier Reef Province are channel-fill deposits, which may reach a thickness in excess of 25 m on the middle and outer shelf. These deposits probably consist of fluviatile sands and muds and may be overlain by estuarine sediments. The infilled channels extend across the shelf to the shelf edge, often coinciding with reef passes. In the Townsville area (Fig. 1), the paleochannels occupied by the Herbert and Burdekin Rivers during the last glacial sea-level low have been mapped across the shelf from the CSP data (Johnson and Searle, in press). Sediment brought to the Coral Sea by these rivers during Pleistocene low-sea-level phases has resulted in 25 km of shelf margin progradation (Symonds et al., 1983). The paleochannels are believed to have been backfilled by fluvial and estuarine deposits, either after channel abandonment or during a rise in base level associated with sea-level rise. (Johnson et al., 1982). Apart from these

channel infill sediments (2% cover), and localized thin and patchy
transgressional sand sheets, postglacial deposits are absent over much of the
middle shelf and inter-reefal areas of the outer shelf (Fig. 11).

6.1.2   Holocene high sea deposits

On the terrigenous-dominated inner shelf, two seismic facies units
corresponding to sedimentation in high and low wave energy environments have
been recognized. Figure 12 illustrates both seismic facies units (Cc and Cf)
offshore from the wave- and tide-dominated subtidal delta of the Herbert River.
The facies unit Cc consists of fluvially derived muddy sands, typically
prograded (foreset gradient 0.05), and may be over 10 m thick in nearshore
shoals. The low-energy depositional facies unit Cf occurs as a mud lobe
overlying the relict sandy veneer Mv. The terrigenous mud and muddy sand facies
are generally restricted to the inner shelf, and thickness of the deposits
increases towards the coast.

In the reef tract, seismic facies units interpreted as Holocene reef
growth and reef talus deposits have been recognized (Fig. 11). Reef talus is
usually 5-10 m thick at the base of reefs, and thins rapidly onto the shelf.
Talus thickness may reach 15 m in the lee of some outer shelf linear reefal
shoals. Over much of the middle shelf and inter-reefal areas of the outer
shelf, the sea floor is mantled by a veneer of mixed carbonate and relict
terrigenous sediment too thin (0.5m) to be resolved by the "Uniboom" CSP system
(Johnson and Searle, in press).

The thickness and relative cover of postglacial terrigenous and carbonate
sediments on the central Great Barrier Reef shelf are shown in Table 1. The
estimates are based on average thicknesses and distribution of lithofacies
interpreted from CSP data in the Townsville region (Johnson and Searle, in
press). Terrigenous sediments clearly dominate, even though the contribution of
coastal progradation is not included in this analysis.

6.2   Preglacial sequences

Limited data are available on the preglacial shelf sequences that underlie
reflector A. Knowledge of the subsurface stratigraphy of the shelf comes from
shallow penetration vibracoring and interpretation of continuous seismic
reflection profiles. These results indicate that modern reefs are underlain by
late Pleistocene reefal platforms, which are commonly more extensive than the
modern reefs (Davies et al., 1981). In addition, the seismic results indicate
that, beneath the outer shelf in the northern region, the pre-Holocene surface

may be an eroded limestone plain (Orme, 1983). Elsewhere in the Reef Province, the pre-Holocene surface on the shelf is underlain by sequences interpreted as coastal plain alluvial, fluvial, and shallow marine deposits (Searle, 1983a,b).

Several erosional discontinuities present in the subsurface presumably resulted from repeated episodes of shelf emergence during the late Quaternary. Fluvial channeling with complex infilling, and prograding fluvial/deltaic facies, is dominant within the pre-Holocene sequences, particularly opposite major drainage systems, and clearly relate to low-sea-level phases. Pre-Holocene sedimentation thus appears to be dominated by fluvial, alluvial and deltaic sedimentation, even though reef carbonate platforms may have been more extensive in the past than they are today.

TABLE 1: The relative area, thickness and volume of postglacial sedimentary facies on the continental shelf off Townsville.

Facies	% cover	Average thickness (m)	% volume
Terrigenous	20	4	55
Mixed carbonate/relict terrigenous	15	1	10
Carbonate	5	10	35
"Nondeposition"	60	–	–

Notes: Based on continuous seismic profiling of 30,300 km^2 of continental shelf between Bowen and Ingham (Johnson and Searle, in press). Areas of nondeposition may include a veneer of modern sediment a few decimetres thick. The area of carbonate deposition includes reef platforms, intrareef lagoons, shoals and patches, and fringing talus deposits, and the thickness quoted is the estimated average of all these. Reef platforms are usually >20 m thick in the central GBR (Davies and Hopley, 1983).

7 QUATERNARY SHELF DEVELOPMENT

Sedimentation in the Great Barrier Reef Province during the Holocene high sea level phase is dominated by the influx of terrigenous sediment along the coast, and by reef framework construction along the outer shelf. The dynamics of the shelf constrains fluvially-derived sediment to water depths of less than 25 metres, and contributes to major coastal progradation. The high turbidity of this inner shelf zone reduces biotic colonization and precludes significant reef construction. Carbonate productivity increases seaward of the coastal turbid zone and is reflected in abrupt changes in bottom sediment facies. Sites of greatest production are represented by reef complexes and by Halimeda bioherms, but there is only limited redistribution of bioclastic detritus away from these structures. Consequently, much of the shelf floor seaward of the 25 m isobath and away from the immediate vicinity of reefs is mantled with only a veneer of reworked relict sediment intermixed with skeletal detritus of the local benthic biota. Only in the northern Great Barrier Reef, where the shelf is narrow and the main reef matrix comes to within 20 km of the coast does significant intermixing of fluvially-derived and reef-derived sediment occur (Flood et al., 1978; Orme and Flood, 1980; Davies and Hughes, 1983).

Under the present sea level framework, shelf aggradation is occurring largely in response to coastal progradation. Strata development occurs as a two layered wedge, with terrigneous shelf muds overlain by a variety of terrigenous intertidal facies, and planated to local supratidal level (Fig. 13). Maximum sedimentation occurs at the coastline, which increases in thickness as it progrades into deeper water. The terrigenous facies thin laterally seaward into a bioclastic carbonate veneer punctuated by individual reef structures. On a uniformly sloping substrate, and with constant sea level and sediment supply, the rate of progradation decreases with time as the coast migrates across the shelf. This is contrary to a model proposed by Chappell et al., (1983) for self-induced accelerating progradation of the coast.

Autochthonous reef growth since the postglacial transgression has occurred at rates of about 30×10^6 tonnes/yr, rates which, over this time scale, are comparable with the terrigenous sediment input of about 28×10^6 tonnes/yr. However, the allochthonous contribution by reefs to the shelf has not been great, and measurements of the modern biodetrital flux on reefs suggest a value very much less than 14×10^6 tonnes/yr and possibly only about 2.5×10^6 tonnes/yr. Rates of in situ skeletal carbonate production on the shelf floor are also thought to be less than 15×10^6 tonnes/yr, and seismic data indicate little or no accumulation over much of the middle shelf and inter-reef areas of the outer shelf. Most reefs have reached or are close to present sea level, so that autochthonous accumulation is expected to decrease in the future. At the same time, reefs may well become increasing contributors of biodetritus to the shelf, whereas the terrigenous input from the mainland is expected to remain essentially constant. This changing role of the reef complexes will be intrinsically dependent upon continuing stability of present sea level.

Glacio-eustatic sea-level fluctuations have characterized the Quaternary (Fig. 14) and rapid sea level fluctuations, rather than sea-level stability, have been the norm. Changing sea levels would have resulted in major changes in the relative contribution of sediment from different sources and in the operative dispersal processes, and consequently would have been of paramount importance in shaping shelf evolution. Although local inundation levels of

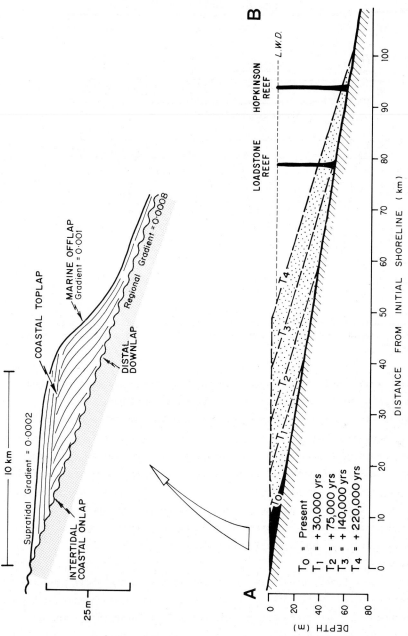

Fig. 13. Present (To) and predicted (To-T4) shoreline progradation across the Townsville continental shelf assuming constant sea level and terrigenous sediment delivery, and the expected seismic stratigraphic elements of the prograding coastal wedge in Bowling Green Bay. Progradation rate decreases as the coast advances across a deepening shelf. Adapted from Belperio (1983a,b).

168

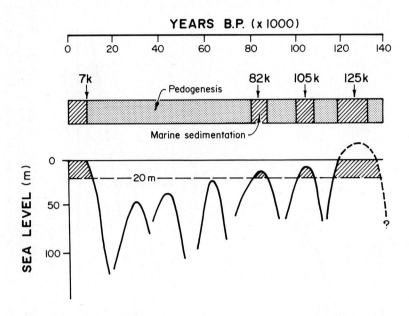

Fig. 14. Late Quaternary sea level curve after Bloom et al. (1974) and Chappell
(1974, 1976), and the anticipated periods of subaerial exposure and pedogenesis
of the inner shelf.

previous high stands are not known with certainty, they would have resulted in
repeated migration of the coastline across the shelf, and hence in a
fluctuating linear terrigenous depocenter. In addition, major regressions
occurring between each high stand would have exposed the entire continental
shelf to pedogenesis, fluvial reworking, and possible alluvial aggradation.
Fluvial input would have continued at all times, irrespective of the position
of the coastline, albeit at different rates and at different across-shelf loci.
A combination of alluviation during low sea-level-phases and terrigenous
coastal sedimentation along a constantly shifting depocenter, suggests that
terrigenous sediments should occur in the subsurface across the entire
continental shelf.

Conversely, carbonate productivity is significant on the shelf today, but
it would have been adversely affected by Quaternary sea-level changes. Holocene
net reef growth is probably the highest since the last interglacial maxima (c.
125,000 years BP), but such periods of reef growth have occurred for only 5 to

10% of the Quaternary (Davies, 1983; Marshall, 1983). During interstadial
sea-level highs, reef growth and carbonate production may have been adversely
affected by increased turbidity associated with a narrower and shallower shelf.
Reef growth and shelf carbonate production would have "switched off" completely
whenever sea level was below about -50 m. In addition, prolonged periods of
subaerial exposure between successive growth phases would have resulted in
significant reef erosion and a reduction in reef thickness (Harvey, 1980;
Marshall, 1983; Davies and Symonds, 1983).

Walker et al. (1983) discussed a model for carbonate to terrigenous
clastic sequences that considers sedimentation rates within adjoining carbonate
and terrigenous depocenters to be essentially equivalent. This assumption is
valid for the Great Barrier Reef Province only under the present sea level
framework, as late Holocene rates of carbonate production and terrigenous input
to the shelf are approximately equal. However, long term sea level changes have
suppressed carbonate production, probably by an order of magnitude, compared to
late Holocene rates, and have resulted in a continual shifting in the
terrigenous depocenter. Models of carbonate to terrigenous sequences must
therefore include relative sea-level change as a prime deterministic parameter.
Transgressions and regressions provide a framework for interdigitation of
terrigenous and carbonate shelf sediments, though preservation of these
sequences may have been affected by subaerial coastal plain processes during
shelf exposure.

Shallow coring and seismic data confirm the predominance of coastal plain,
alluvial, fluvial, and shallow marine terrigenous deposits throughout the
subsurface of the shelf (Searle, 1983a; Davies et al., 1983). The reef
complexes are largely isolated, multiple-generation, carbonate structures that
initially grew on, and occur within, siliciclastic, fluviatile and deltaic
sediments (Davies, 1983; Symonds, 1983). Intermediate and deep-penetration
seismic data confirm the long-term importance of terrigenous sedimentation in
shelf-margin evolution, and continued subsidence of the shelf has resulted in a
stacked sequence of progradational units dominated by terrigenous sediments
(Symonds, 1983; Symonds et al., 1983).

8 ACKNOWLEDGEMENTS

A.P. Belperio publishes with permission of the Director-General,
Department of Mines and Energy, South Australia. D.E. Searle publishes with
permission of the Chief Government Geologist, Department of Mines, Queensland.

We thank N. Harvey for valuable critical comments. Figures 1, 3, 10, 11 and 12 are reproduced by permission of the Director, Bureau of Mineral Resources, Geology and Geophysics, Canberra, Australia.

9 REFERENCES

Arnold, P.W., 1980. Soft-bottom polychaete communities in Halifax Bay, Queensland. Ph.D. Dissertation, James Cook University, Townsville, Queensland, 180 pp.

Belperio, A.P., 1978. An inner shelf sedimentation model for the Townsville region, Great Barrier Reef Province. Ph.D. Dissertation, James Cook University, Townsville, Queensland, 210 pp.

_____ 1979a. Negative evidence for a mid-Holocene high sea level along the coastal plain of the Great Barrier Reef Province. Mar. Geol., 32. M1-M9.

_____ 1979b. The combined use of wash load and bed material load rating curves for the calculation of total load; an example from the Burdekin River, Australia. Catena, 6, 317-329.

_____ 1983a. Terrigenous sedimentation in the central Great Barrier Reef lagoon; a model from the Burdekin region. BMR Jour. Australian Geol.   and Geophys., 8, 179-190.

_____ 1983b. Late Quaternary terrigenous sedimentation in the Great Barrier Reef Lagoon. In: J.T. Baker, R.M. Carter, P.W. Sammarco, and K.P. Stark (Editors), Proc. Inaugural Great Barrier Reef Conf. Townsville, Queensland, JCU Press, pp. 71-76.

_____ McMANUS, J. AND COHEN, P.H., 1983, A Fortran program for suspended sediment dynamics and tidal flux monitoring: Computers and Geosciences, 9, 221-227.

Birtles, A. and Arnold, P., 1983. Between the reefs; some patterns of soft substrate epibenthos on the central Great Barrier Reef Shelf. In: J.T. Baker, R.M. Carter, P.W. Sammarco, and K.P. Stark (Editors), Proc. Inaugural Great Barrier Reef Conf., Townsville, Queensland, JCU Press, pp. 159-163.

Bloom, A.L., Broecker, W.S., Chappell, J.M.A., Mathews, R.K., AND Mesolella, K.J., 1974. Quaternary sea level fluctuations on a tectonic coast; new 230Th/234U dates from the Huon Peninsula, New Guinea, Quaternary Research 4, 185-205.

Burgis, W.A., 1974. Cenozoic history of the Torrilla Peninsula, Broad Sound, Queensland. Bureau of Mineral Resources, Australia, Report 172, 42.

Chappell, J., 1974. Geology of coral terraces, Huon Peninsula, New Guinea; a study of Quaternary tectonic movements and sea level changes. Geol. Soc. America Bull., 85, 533-570.

_____ 1976. Aspects of late Quaternary palaeogeography of the Australian-east Indonesian region. In: R.L. Kirk, and A.G. Thorne, (Editors),   The Origin of the Australians: Canberra, Australian Institute of Aboriginal Studies, pp. 11-22.

_____ 1983. Evidence for smoothly falling sea level relative to north Queensland, Australia, during the past 6000 years. Nature 302, 406-408.

_____ Chivas, A., Wallensky, E., Polach, H.A., AND Aharon, P., 1983. Holocene palaeo-environmental changes, central to north Great Barrier Reef inner zone. BMR Jour. Australian Geol. and Geophys. 8, 223-235.

Chivas, A.R., Torgersen, T., and Andrews, A.S., 1983. Isotopic tracers of Recent sedimentary environments in the Great Barrier Reef. In: J.T. Baker,   R.M. Carter, P.W. Sammarco, and K.P. Stark (Editors), Proc. Inaugural Great Barrier Reef Conf., Townsville, Queensland, JCU Press, pp. 83-88.

Cook, P.J., AND Mayo, W., 1978. Sedimentology and Holocene history of a tropical estuary (Broad Sound, Queensland). Bureau of Mineral Resources, Australia, Bull. 170, 206 pp.

Davies, P.J., 1977. Modern reef growth, Great Barrier Reef: Proc. 3rd Intern. Symp. Coral Reefs, Miami, 2, pp. 325-330.

_____ 1983. Geo-reflections on the Great Barrier Reef. In: J.T. Baker, R.M. Carter, P.W. Sammarco, and K.P. Stark (Editors), Proc. Inaugural Great Barrier Reef Conf., Townsville, Queensland JCU Press, pp. 13-25.

_____ and Hopley, D., 1983. Growth facies and growth rates of Holocene reefs in the Great Barrier Reef. BMR Jour. Australian Geol. and Geophys., 8, 237-251.

_____ and Hughes, H., 1983. High-energy reef and terrigenous sedimentation, Boulder Reef, Great Barrier Reef. BMR Jour. Australian Geol. and Geophys., 8, 201-209.

_____ and Kinsey, D.W.,, 1977. Holocene reef growth, One Tree Island, Great Barrier Reef. Mar. Geol., 24, 1-11.

_____ and Symonds, P.A., 1983. Evolution of the Great Barrier Reef Province. Bureau of Mineral Resources, Australia, Record 1983/4, 5-6.

_____ and West, B.G., 1981. Suspended sediment transport and water movement at One Tree Reef, southern Great Barrier Reef. BMR Jour. Australian Geol. and Geophys., 6, 187-195.

_____ Marshall, J.F., Hekel, H., and Searle, D.E., 1981. Shallow inter-reefal structure of the Capricorn Group, southern Great Barrier Reef. BMR Jour. Australian Geol. and Geophys., 6, 101-105.

_____ Cucuzza, J., and Marshall, J.F., 1983. Lithofacies variations on the continental shelf, east of Townsville, Great Barrier Reef. In: J.T. Baker, R.M. Carter, P.W. Sammarco, and K.P. Stark (Editors), Proc. Inaugural Great Barrier Reef Conf., Townsville, Queensland, JCU Press, pp. 89-93.

DeKeyser, F., 1964. Innisfail, Queensland 1:250 000 Geological Series. Bureau of Mineral Resources, Australia, Explanatory Notes SE/55-6, 30 pp.

_____ Fardon, R.S.H., AND Cuttler, L.G., 1965. Ingham, Queensland 1:250 000 Geological Series: Bureau of Mineral Resources, Australia, Explanatory Notes SE/55-10, 31 pp.

Dowling, J.J., 1977. A grain size spectral map. Jour. Sedimentary Petrology, 47, 281-284.

Drew, E.A, 1983. Halimeda biomass, growth rates and sediment generation on reefs in the central Great Barrier Reef Province. Coral Reefs, 2, 101-110.

Falvey, D.A., and Mutter, J.C., 1981. Regional plate tectonics and the evolution of Australia's passive continental margins. BMR Jour. Australian Geol. and Geophys., 6, 1-29.

Flood, P.G., and Orme, G.R., 1983. Mixed siliciclastic/carbonate shelf deposits of the northern Great Barrier Reef province. In: J.T. Baker, R.M. Carter, P.W. Sammarco, and K.P. Stark (Editors), Proc. Inaugural Great Barrier Reef Conf., Townsville, Queensland, JCU Press, pp 105.

_____, _____ and Scoffin T.P., 1978. An analysis of the textural variability displayed by inter-reef sediments of the impure carbonate facies in the vicinity of the Howick Group. Phil. Trans. Royal Soc. London, 291A, 73-83.

Frankel, E., 1971. Recent sedimentation in the Princess Charlotte Bay and Edgecumbe Bay areas, Great Barrier Reef Province. Ph.D. Dissertation, University of Sydney, Sydney, New South Wales.

_____ 1974. Recent sedimentation in the Princess Charlotte Bay area, Great Barrier Reef Province. Proc. 2nd Intern. Symp. Coral Reefs, Brisbane, v.2, p. 355-369.

Frith, C.A., 1983. Aspects of lagoon sedimentation and circulation at One Tree Reef, southern Great Barrier Reef. BMR Jour. Australian Geol. and Geophys., 4. 8, p. 211-221.

Harvey, N., 1980. Seismic investigations of a pre-Holocene substrate beneath modern reefs in the Great Barrier Reef Province. Ph.D. Dissertation, James Cook University, Townsville, Queensland, 329 pp.
_____ and Hopley, D., 1981. The relationship between modern reef morphology and a pre-Holocene substrate in the Great Barrier Reef Province. Proc. 4th Intern. Symp. Coral Reefs, Manila, 1, 549-554.
_____ Davies, P.J., and Marshall, J.F., 1979. Seismic refraction - a tool for studying coral reef growth. BMR Jour. Australian Geol. and Geophys., 4, 141-147.
Hopley, D., 1983. Evidence of 15,000 years of sea level change in tropical Queensland. In: Hopley, D., (Editor), Australian Sea Levels in the Last 15,000 Years, a Review. James Cook University of North Queensland, Department of Geography Monograph Series, Occasional Paper, 3, 93-104.
Johnson, D.P., and Searle, D.E., in press. Post-glacial seismic stratigraphy, central Great Barrier Reef, Australia: Sedimentology.
_____, _____ and Hopley, D., 1982. Positive relief over buried post-glacial channels, Great Barrier Reef Province, Australia. Mar. Geol., 46, 149-159.
Jones, M.R., and Stephens, A.W., 1983. Transport of beach sand around headlands, Trinity Bay, North Queensland; Implications for coastal change. Queensland Government Mining Jour., v. 84, p. 5-11.
Kinsey, D.W., 1981. The Pacific/Atlantic reef growth controversy. Proc. 4th Intern. Symp. Coral Reefs, Manila 1, p. 493-498.
_____ and Davies, P.J., 1979. Inorganic carbon turnover, calcification and growth in coral reefs. In: P.A. Trudinger, D.J. Swaine (Editors), Biogeochemical Cycling of Mineral Forming Elements. Amsterdam, Elsevier, pp. 131-162.
Marshall, J.F., 1977. Marine geology of the Capricorn Channel area. Bureau of Mineral Resources, Australia, Bull. 163, 81 pp.
_____ 1983. The Pleistocene foundation of the Great Barrier Reef. In: J.T. Baker, R.M. Carter, P.W. Sammarco, and K.P. Stark (Editors), Proc. Inaugural Great Barrier Reef Conf., Townsville, JCU Press, pp. 123-128.
_____ and Davies, P.J., 1978. Skeletal carbonate variation on the continental shelf of eastern Australia. BMR Jour. Australian Geol. and Geophys., 3, 85-92.
_____ and _____ 1982. Internal structure and Holocene evolution of One Tree Reef, Southern Great Barrier Reef. Coral Reefs, v. 1, p. 21-28.
Maxwell, W.G.H., 1968. Atlas of the Great Barrier Reef. Amsterdam, Elsevier, 258 p.
_____ and Swinchatt, J.P., 1970. Great Barrier Reef; Regional variation in a terrigenous-carbonate province. Geol. Soc. America Bull., 81, 691-724.
Mutter, J.C., and Karner, G., 1980. The continental margin off northeast Australia. In: R.A. Henderson, and P.J. Stephenson (Editors), The Geology and Geophysics of Northeastern Australia. Brisbane, Geological Society of Australia (Queensland Division), pp. 47-69.
Ollier, C.D, 1978. Tectonics and geomorphology of the eastern highlands. In: J.L. Davies, and M.A.J. Williams (Editors), Landform Evolution in Australasia. Canberra, ANU Press, pp. 5-47.
Orme, G.R., 1983. Shallow structure and lithofacies of the northern Great Barrier Reef. In: J.T. Baker, R.M. Carter, P.W. Sammarco, and K.P. Stark (Editors), Proc. Inaugural Great Barrier Reef Conf., Townsville, JCU Press, pp. 135-141.
_____ and Flood, P.G., 1977. The geological history of the Great Barrier Reef; a reappraisal of some aspects in the light of new evidence. Proc. 3rd Intern. Symp. Coral Reefs, Miami, v. 2, p. 37-43.

_____ and _____ 1980. Sedimentation in the Great Barrier Reef Province, adjacent bays and estuaries. In: R.A. Henderson, and P.J. Stephenson (Editors), The Geology and Geophysics of Northeastern Australia: Brisbane, Geological Society of Australia (Queensland Division), p. 419-434.

_____ Webb, J.P., Kelland, N.C., and Sargent, G.E.G., 1978a. Aspects of the geological history and structure of the northern Great Barrier Reef. Phil. Trans. Royal Soc. London, 291A, 23-35.

_____ Flood, P.G., and Sargent, G.E.G., 1978b. sedimentation in the lee of outer (ribbon) reefs, Northern Region of the Great Barrier Reef Province. Phil. Tans. Royal Soc. London, 291A, 85-99.

Paine, A.G.L., 1972. Proserpine, Queensland 1:250 000 Geological Series. Bureau of Mineral Resources, Australia, Explanatory Notes SF/55-4, 24.

Pickard, G.L., 1977. A review of the physical oceanography of the Great Barrier Reef and Western Coral Sea. Australian Institute of Marine Science, Monograph Series, 2, 135.

Pringle, A.W., 1983. Sand spit and bar development along the east Burdekin Delta coast, Queensland, Australia. James Cook University of North Queensland, Department of Geography Monograph Series, 12, 34.

Searle, D.E., 1979. Trinity Bay continuous seismic profiling survey. Geological Survey of Queensland, Record 1979/1. 25 p.

_____ 1983a. Late Quaternary regional controls on the development of the Great Barrrier Reef; geophysical evidence. BMR Jour. Australian Geol. and Geophys., 8, 267-276.

_____ 1983b. Shallow seismic structure-southern reefs. In: J.T. Baker, R.M. Carter, P.W. Sammarco, and K.P. Stark (Editors), Proc. Inaugural Great Barrier Reef Conf., Townsville, JCU Press, 143-149.

_____ and Hegarty, R.A., 1982. Results of a continuous seismic profiling survey in the Princess Charlotte Bay area. Geological Survey of Queensland, Record 1982/17, 25 p.

_____ Harvey, N., and Hopley, D., 1980. Preliminary results of continuous seismic profiling in the Great Barrier Reef Province between 16°10'S and 19°20'S. Geological Survey of Queensland, Record 1980/23, 32p.

_____, _____, _____ and Johnson, DP.P., 1981. Significance of results of shallow seismic research in the Great Barrier Reef Province between 16°10'S and 20°05'S. Proc. 4th Intern. Symp. Coral Reefs, Manila 1, 531-539.

Smith, S.V., and Kinsey, D.W., 1976. Calcium carbonate production, coral reef growth, and sea level change. Science, 194, 937-939.

Symonds, P.A., 1983. Relation between continental shelf and margin development - central and northern Great Barrier Reef. In: J.T. Baker, R.M. Carter, P.W. Sammarco, and W.P. Stark (Editors), Proc. Inaugural Great Barrier Reef Conf., Townsville, Queensland, JCU Press, pp. 151-157.

_____ Davies, P.J., and Parisi, A., 1983. Structure and stratigraphy of the Great Barrier Reef. BMR Jour. Australian Geol. and Geophys., 8, 277-291.

Taylor, L.W.H., and Falvey, D.A., 1977. Queensland Plateau and Coral Sea Basin, stratigraphy structure and tectonics. The APEA Jour., 17, 13-29.

Thom, B.G., and Roy, P.S., 1983. Sea level change in New South Wales over the past 15,000 years. In: D. Hopley (Editor), Australian Sea Levels in the last 15,000 years, A Review: James Cook University of North Queensland, Department of Geography Monograph Series, occasional Paper, 3, 64-84.

Torgersen, T., Chivas, A.R., and Chapman, A., 1983. Chemical and isotopic characterisation and sedimentation rates in Princess Charlotte Bay, Queensland. BMR Jour. Australian Geol. and Geophys., 8, 191-200.

174

Walker, T.A., 1981a. Dependence of phytoplankton chlorophyll on bottom resuspension in Cleveland Bay, northern Queensland. Australian Jour. Mar. Freshwater Res., 32, 981-986.

_____ 1981b. Annual temperature cycle in Cleveland Bay, Great Barrier Reef Province. Australian Jour. Mar. Freshwater Res., 32, 987-991.

Walker, K.R., Shanmugam, G., and Ruppel, S.C., 1983. A model for carbonate to terrigenous clastic sequences. Geol. Soc. America Bull., 94, 700-712.

Wolanski, E., and Jones, M., 1981. Physical properties of Great Barrier Reef lagoon waters near Townsville, I. Effects of Burdekin River Floods. Australian Jour. Mar. Freshwater, Res., 32, 305-319.

Chapter 6

MIXED SILICICLASTIC/CARBONATE SEDIMENTS OF THE NORTHERN GREAT BARRIER REEF
PROVINCE, AUSTRALIA

P.G. FLOOD and G.R. ORME
Department of Geology, University of New England, Armidale, NSW; Department
of Geology, University of Queensland, St. Lucia, Qld.

ABSTRACT
    Lithofacies maps of the northern Great Barrier Reef Province based
merely on the proportions of carbonate and siliciclastic components in
surface sediments present an over-simplified picture of an across-the-shelf
carbonate source. The decrease in the proportion of siliciclastic components
across the shelf is not gradual because of the dominance of along-the-shelf
sediment transport in the near-shore and inner-shelf. Reefs, in situ
post-mortem contributions, and Halimeda meadows are the main sources of
carbonate end-members. Siliciclastic end-members are derived from the
mainland and high-islands. Mixing of end-members is an important process in
determining the nature of most inter-reef sediments, while others may
represent lag deposits, dependent for their grain size characteristics upon
in situ post-mortem contributions and currents of removal. Cores from the
outer shelf banks that are veneered by present Halimeda meadows indicate
that Halimeda has been an important source of carbonate in the area during
the Holocene.

1 INTRODUCTION

    The generalized lithofacies maps of shelf sediments produced by earlier
researchers (Maxwell, 1968, 1973; Maxwell and Swinchatt, 1970; see also
Ginsburg and James, 1974) were based on the proportion of carbonate and
terrigenous (non-carbonate) end-members. It was suggested that such
end-members reflect the dominance of the western, mainland terrigenous
provenance, and the outer-shelf (reefal) carbonate source on the east, with
a central zone which shows the influence from either source (Maxwell and
Swinchatt, 1970). Consequently it was thought that the shelf-floor sediments
showed a semi-meridional zonation of lithofacies with near shore and inner
shelf of terrigenous sediments, a zone of carbonate sediments on the outer
shelf, and a mid-shelf zone which is dominated in places by sediments of
varied composition. It was also suggested that 'across the shelf
transportation' of terrigenous sediment is "aided by regional shoaling of
the shelf toward the north" and by the inferred "more effective transporting
action of bottom currents".

    Recent studies, however, have shown that transportation of terrigenous
sediment beyond the inner shelf is limited, and with the exception of some
occasional diffusion of suspended sediment, transportation of terrigenous

sediment delivered to the near shore zone is along the shelf rather than across it (Belperio, 1983). Zones of rapid lateral accretion are widespread along the coastline especially in the vicinity of the larger rivers. The clay and very fine silt delivered to the shelf by rivers become flocculated and may be deposited to form part of a prograding wedge extending some distance from the shoreline. The northward transport of flocculated clays and fine silt from the Barron and Endeavour Rivers and the northward movement of quartz sand by longshore drift has been noted by previous researchers.

On the narrow northern shelf high-carbonate sediments of reefal origin are limited to narrow peripheral zones extending only 1km to 3km from the major reefs (Flood et al., 1978). Also on the shelf near Lizard Island, more extensive areas of high carbonate sediment are due to factors other than direct reefal influence (Orme et al., 1978). Extensive Halimeda banks are developed landward of the outer barrier reefs.

The purpose of this paper is to review the evolution of ideas on the development of Carbonate to Clastic Facies Changes within the northern Great Barrier Reef (Fig. 1). In addition, the significance of in situ post-mortem contribution of carbonate to the mid shelf sediments will be highlighted, especially the influence of Halimeda banks and benthic foraminiferans.

2 GENERAL CHARACTERISTICS OF THE AREA

The morphology of the northern Great Barrier Reef Province conforms to the classical model of a rimmed shelf. Barrier (ribbon) reefs extend along 70% of the shelf edge, the shelf is generally less than 40m deep and varies in width from between 30 to 50km. A longitudinal zonation is developed across the shelf. The following zones have been described by Maxwell (1968): near shore (0-10m); inner shelf (10-40m); and marginal shelf (40-100m).

Reef distribution in the lagoon behind the shelf edge is also variable. There tends to be a fairly uniform dispersion of platform reefs over the outer two-thirds of the shelf, and well developed fringing reefs occur around the high continental islands.

Four major zones of reef development may be recognized:

1. A line of ribbon reefs, with narrow intervening passages situated on the outer edge of the shelf. The southern limit of these is just north of Cairns.

2. A mid-shelf zone of reefs often with narrow intervening passages. In the same zone exists a number of smaller irregular spaced reefs separated from others by up to 8km of open water.

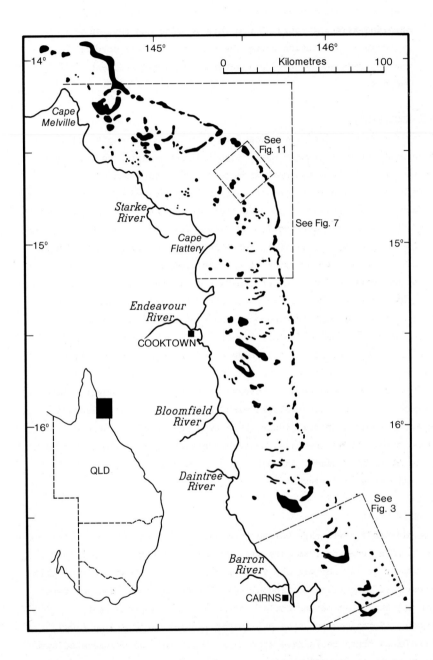

Fig. 1.   Distribution of reef (black) along the shelf in the northern Great
Barier Reef.

3. An inner-shelf zone of low wooded island reefs and small
   reefs with sand cays generally within 20km of the coast.
4. Mainland fringing reefs. These vary in size and have an
   irregular distribution along the coastline. Their development is related to the local water quality and sediment
   budgets.

3 SHELF-EDGE REEFS

A chain of linear reefs, varying from 3 to 25 km in length and 300 to
450 m in width occur along the shelf edge aligned parallel to the shelf
margin so that they have maximum exposure to the open ocean with its
westward moving currents and the oceanic swell generated by the prevailing
South East Trades. They are separated from each other by narrow passages.
Day Reef, Carter Reef, Yonge Reef, and Ribbon Reef to the north and east of
Lizard Island are typical of this reef type. Their eastward extension is
restricted by the deepening floor of the continental slope, their westward
growth by the partial isolation of the back-reef areas from the open oceanic
waters. The growth of the ends is checked at the narrow channels where
successive reefs approach one another, and continued growth occurs only
where the bathymetry and hydrology are favorable, which results in a
backward projection of reef ends. These cuspate projections modify the
current patterns in and around the lee of the reefs so that lateral
extensions (prongs) develop and the cusps become recurved. Small triangular
reefs (plug reefs) may be developed with their apex pointed oceanward
through the openings in the linear reefs as a result of the strong tidal
currents through these openings. Because the ribbon reefs are in such a high
energy environment sediment is unable to accumulate and the surface is a
swept, flat, cemented pavement.

4 MID-SHELF REEFS

The mid shelf reefs have varied form and size. From 16° to 17° the reefs
are up to 26km long and 12km wide with crescentic fronts facing into the SE
trades. North of 16° the mid shelf reefs are smaller. These reefs usually
only have hard line (i.e. continuous) development on their southeastern
margins and their perimeters enclose areas of complex patch reef
development. Sometimes the crescent shaped margins enclose deep (20m)
lagoons. Often these reefs have sand cays (vegetated and unvegetated types)
developed on their leeward margin.

Small ephemeral cays or intertidal sand patches are developed on some
reefs whereas on others vegetated cays with low shrub development are
common.

5 INNER-SHELF REEFS

The low wooded island reef type occur on the inner shelf and is
restricted to the sector of the continental shelf between 12°S and 17°S. It
usually occurs within 15km of the coast and from 25 to 40km from the outer
reefs of the Province. Several models have been proposed to explain the
limited occurrence of this reef type. Stoddard (1965) who studied similar
features of Atlantic reefs suggested that the location and distribution of
the low wooded island reef types conform to a simple energy model in which
the degree of exposure to wave action is a major control. 'The factors
implicit in exposure in this context include variability and strength of
wind, effective fetch, depth of water to windward, and local factors of reef
depth and geometry'. In summary, the energy model suggests:

1. The outermost linear reefs are situated in the maximum
   energy area. Rates of reef growth and sediment production
   are high but wave energy is sufficient to carry all the
   debris across the reef to the back-reef area. Single ramparts
   and cays are not formed.
2. The central mid-shelf reefs are situated in a medium-energy
   area. Rate of reef growth is high, but shingle does not
   accumulate above sea level because of the weaker wave
   action; sand production is high and sand cays are formed by
   wave refraction.
3. The low wooded island reef types are located in areas of
   moderate to low energy level. Reef growth consists of
   numerous fragile species which are available for fragmentation
   and shingle production during major storms because the fetch
   and water depth to windward are adequate for destructive wave
   development. Sediment supply is high and energy is sufficient
   to lift the debris onto the roof edge, but it is not strong
   enough to transport it across the reef flat. Shingle
   accumulates on the windward part of the reef and sand is
   concentrated by wave refraction towards the lee of the reef
   flat. The lower-energy environment protected by the shingle
   rampart may be subsequently colonized by mangroves.

6 FRINGING REEFS

The near-shore reefs rising from shallow depths (10m) include a variety of fringing and barrier types in terms of their back-reef character.

The fringing reefs of the continental islands such as Lizard Island and North and South Direction Islands contrast with those of the mainland coast in that coral growth is much more vigorous and varied in the less turbid waters. Lizard Island is almost completely surrounded by well developed fringing reefs. Along the steep eastern and northeastern coastline they are rarely more than 50m in width, rising steeply from depths of -20m. A more extensive reef is developed between Lizard Island and the small islets to the south enclosing a fairly deep lagoon (10m) having an open entrance to the west alongside Lizard Island.

7 CONTINENTAL ISLANDS

Lizard Island (14°14'S, 145°28'E) is a high rocky granitic island 24km from the mainland and 16km from the shelf-edge reefs. The island has an area of approximately $10km^2$ and a maximum N-S length of 45km; E-W width is 3km. Two NNW-trending ridges cross the island separated by a low grassy valley; the highest point of the island is 360m. Two smaller rocky islets, South Islet and Palfrey Islet are 2km SW and 2.5km WSW of the southeastern extremity of Lizard Island, respectively. Lizard Island is bordered by narrow, steep fringing reefs on its eastern and western sides. To the south the reefs are more extensive and join with the reefs fringing the southern islets to enclose a small lagoon up to 9m in depth. Sandy quartz beaches occur in the bays and inlets of the island.

8 HALIMEDA BANKS

On the outer shelf near Lizard Island special circumstances have promoted the luxuriant growth of 'Halimeda meadows' and the pure carbonate lithofacies is here dominated by Halimeda debris (gravels and coarse sands) which give rise to a distinct subfacies. These banks are up to 18m thick and have a fairly uniform depth of -25m. Orme et al. (1978) attribute this to luxuriant Halimeda growth and sea level control of sediment accumulation in a tidal environment. The seismic reflector beneath these banks in the Lizard Island area extends to the back reef of Carter Reef.

9 MAINLAND

Three geomorphic units may be recognised between Cairns and Lizard Island.

1. The Barron River Deltaic Plain

The deltaic plain and adjacent areas around Cairns contain deltaic deposits which are at least 40m thick. The surface formation consists of up to 4m of quartzose sand with occasional pebbles and shells, underlain by a soft blue-grey clay formation of mangrove muds. The soft clays thicken southwards, reaching a depth of 24m beneath Cairns harbour. Underneath them is a firm yellow-grey clay up to 12m thick which in turn rests on a basement of sandy gravel. Since Holocene sea level first rose to its present position about 6000 years ago, the Barron River has been building up a deltaic sequence much of which consists of beach ridges and associated units.

2. The Coastal Range

This extends from the Barron Delta to the Daintree River. The seaward slopes of the range consist mainly of piedmont aprons which have been modified by marine processes. The coastline is a series of exposed, weathering headlands, bayhead beaches of sands derived from the reworked piedmont fans, and some intervening areas of boulder and shingle beaches, the coarser residual materials from the same piedmont aprons. Coral reefs are developed sporadically close inshore and small fringing reefs are found off some rocky shores. The best known is that at Yule Point (Bird, 1971). It is a former patch reef joined to the mainland by sedimentation. The lower part of this reef is related to a sea level of at least 1m above present and has been dated at 4130 ± 110 years B.P.

3. The Sand Dune Complexes north of Cooktown

The most impressive parabolic dunes on the tropical Queensland coast occur along the eastern seaboard north of Cooktown, most notably in the Cape Flattery areas. This dune field, covering an area of 700km^2 consists largely of elongate parabolic dunes up to 5km in length and over 100m in height. Many of the dunes are stabilized beneath heath, scrub or vine forest but up to 15% of the Cape Flattery dune field consists of active dunes and a further 10% comprises swamps and lakes enclosed between the trailing arms of the dunes. The direction of movement of the dunes from south-east to north-west is parallel to the prevailing wind direction.

10 PREVIOUS WORK

Results of investigations of the shelf sediments in the northern Great Barrier Reef were first published by Maxwell (1968) who on the basis of the carbonate content (acid soluble component) classified the shelf sediments as

High Carbonate (more than 80% soluble), Impure-Carbonate (60-80% soluble),
Transitional (40-60% soluble), Terrigenous (20-40% soluble), and a locally
developed High-Terrigenous Facies (less than 20% soluble). Subsequently
Maxwell and Swinchatt (1970) and Maxwell (1973) provided a more
comprehensive picture of regional differences in the facies patterns. Their
findings were summarized in the work of Ginsburg and James (1974, Fig. 16).

More recently, results of the Royal Society of London and Universities
of Queensland Expedition to the northern Great Barrier Reef (see Flood et
al., 1978; Orme et al., 1978) have greatly modified (see Fig. 2) the earlier
held belief that the lithofacies on the shelf merely reflected the dominance
of a mainland terrigenous source and an outer-shelf (reefal) carbonate
source, with a central zone which showed variable influence of both sources.
Although these new findings have been incorporated in a recent review paper
by Orme and Flood (1980) the evidence for such conclusions will be outlined
here for completeness.

11 CONSTITUENTS OF SHELF-FLOOR SEDIMENTS

Quartz, clay minerals, foraminifera, molluscs, Halimeda, bryozoans, and
corals are the main components of shelf-floor sediments.

12 INORGANIC COMPONENTS

Although the general distribution of inorganic constituents of the
shelf-floor sediments is known, little detailed information regarding their
petrology is available. In addition to quartz and clay minerals, rock
fragments, feldspars, heavy minerals, glauconite, and mica locally reach
significant proportions.

Quartz.- The principal areas of the main quartzose facies (greater than
60% quartz) have been defined and appear to be related to discharge from
major river systems. In a zone extending between Cooktown and Cape Melville,
which is narrow (4km) due to the narrowness of the shelf and the close
proximity of reefs, carbonate effectively dilutes the quartz contribution.
Very little quartz occurs in the carbonate sediment of the outer part of the
shelf except adjacent to continental islands. It is possible that the
Endeavor, Starcke, and other Rivers which drain sandy alluvia, still deliver
large quantities of sand to the shelf but no quantitative investigation of
sediment transportation loads has been made in this area. In addition local
quartz-rich facies occur in the lee of continental islands such as Lizard
Island.

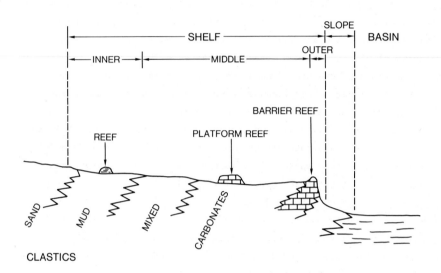

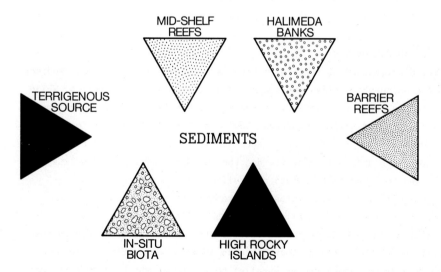

Fig. 2.   Conceptual model of the possible siliciclastic (dark) and
carbonate (light) contributions to the shelf sediments in the
northern Great Barrier Reef.

Clay minerals.- Flocculated clay particles are the major component of the terrigenous muds, and are also significant in the transitional and impure carbonate facies. They are derived from the reworking of argillites and phyllites, or from the chemical weathering of a variety of sedimentary, metamophic and igneous rock types.

Lithic fragments.- Rock fragments occur locally in gravels and sands near the mainland shore and adjacent to continental islands. Elsewhere on the shelf-floor they are associated with relict deposits, or are located in the vicinity of off-reef slopes. They may consist of material of mainland or continental island provenance (such as phyllite, schist, metagreywacke, granite, chert) or may be composed of younger carbonate material in the form of beackrock, cayrock, rampart rock or reef rock.

Ooids.- Diagenetically unaltered aragonite ooids with radial fabric and nuclei of quartz and carbonate grains have been described from Lizard Island where they may constitute up to 10% of channel sands lying between Lizard and Palfrey Islands (Davies and Martin, 1976; Flood, 1983).

Authigenic minerals.- The products of authigenesis are locally conspicuous. Glauconite may be associated with foraminiferal tests in the more terrigenous facies of the near shore and inner shelf. The reduction of iron to the sulphide due to reducing conditions in the bottom sediment may cause black staining of foraminiferal tests which gives rise to the 'Black Speckled Sands' of the inner shelf; oxidation of the iron sulphide will give rise to 'Rusty Sands' or 'Brown Speckled Sands'.

13 SKELETAL COMPONENTS

The distribution of skeletal carbonate allochems in shelf-floor sediment varies considerably across the shelf and also from region to region. While some are obviously associated with the reef tract (for example corraline algae, corals and Halimeda) others show no such affinity and are merely diluted by reef-derived sediment. Reef density, shelf width and bathymetry, tidal currents and surface drift, and the influence of relict deposits are considered to be salient factors controlling allochem distribution.

Foraminifera.- Foraminiferans are probably the most abundant and widespread of the skeletal allochems of shelf-floor sediment. Their distribution pattern differs from that of other skeletal allochems in that benthic and pelagic species contribute as well as reef derived allochems. There are strong facies trends parallel to the coast which reflect the main benthic component. Transportation of pelagic foraminifera across the narrow shelf of the Northern Region has resulted in their concentration in shallow,

nearshore mud. High values for the foraminiferal component (up to 60%) of
inter-reef areas are characteristic. Access of oceanic waters, tidal flow,
surface drift, lack of dilution by reef debris, and the existence of
particularly favorable benthic environments are factors that determine the
distribution of foraminifera in shelf-floor sediment. The common types
include <u>Alveolinella</u>, <u>Marginopora</u>, <u>Operculina</u> and <u>Elphidium</u>.

   <u>Molluscs</u>.- Molluscan debris constitutes the second most abundant and
widespread skeletal allochem. The greater abundance of this component may
indicate less dilution by reef debris, the presence of relict shell
deposits, and the narrow, shallower sandy nature of the shelf.

   <u>Echinoderms</u>.- Echinoderm debris exceeds 5% only in the mud and muddy
sandy areas facies of the inner shelf.

   <u>Halimeda</u>.- Although the reefs do provide some contribution the
significant proportion of <u>Halimeda</u> debris occurs in sediment associated with
submarine banks; values of 70%-90% for inter-reef sediments of the outer
shelf are not unusual, for example in the area on the shelf adjacent to
Lizard Island (Orme <u>et al</u>., 1978). On the western third of the shelf the
content of <u>Halimeda</u> debris falls below 1%.

   <u>Coral</u>.- A high coral content is regarded as indicative of reefal
influence.

   <u>Coralline Algae</u>.-The distribution of this allochem type closely follows
the reef tract, where it may constitute over 5% of the shelf-floor sediment
near to reefs. Its abundance is dependent on reef density.

   Components of shelf-floor sediments have six possible origins:
reef-derived carbonate, <u>in situ</u> post-mortem skeletal carbonate, mainland
terrigenous muds and sands, terrigenous contributions from continental
islands, relict shelf accumulations, and locally significant contributions
of windblown sand.

14 SELECTED AREAS

   Sediment characteristics and distribution patterns are determined by
complex interaction between mainland and continental island sources, and
ecological controls which determine reef development and which promote
conditions favorable to maintenance of luxuriant <u>Halimeda</u> growth over the
outer shelf. Specific examples illustrating the nature and distribution
patterns of the sediments will be discussed.

15   SHELF SEDIMENTS NEAR CAIRNS

Maxwell and Swinchatt (1970) showed that the shelf sediments near Cairns
(Figs. 3-5) consisted of a narrow coastal belt of terrigenous sands,
extending seaward, in most instances, for less than 2km. For the next 4-20km
seaward there occurs a prograding wedge of dominantly terrigenous mud.
Further seaward the mud content decreases as the calcium carbonate gravel
and sand contribution increases in proximity to the reefs. The transition
from 70% to 25% mud content in the sediments may take place within a 2km
distance. A relatively abrupt break-in-slope is associated with the
transition zone between terrigenous and reef derived sediments.

The spatial distribution of the lithofacies (Fig. 6) on the shelf near
Cairns reflects the dominance of the western, mainland terrigenous
provenance, an outer shelf (reefal) carbonate source to the east, and a
central zone which shows the influence of mixing of sediment from either
source. In some areas the central shelf shows little influence of either
source and appreciable sediment accumulation are mainly relict.

Major inter-reef channels are floored by fine-grained, mixed
terrigenous-carbonate sediments. Silt- and clay-sized material, comprising
45% to 70% of the channel sediments, increases toward the channel axes and
also increases seaward. Abundance of acid-insoluble material is fairly
uniform along the length of the channels decreasing only slightly seaward.
Siliceous sponge spicules and other acid-insoluble micro-organisms replace
the disappearing silt- and clay-sized terrigenous material in the
calculation of total insolubles.

The sand fraction of the channel sediments is fine to very fine and
consists of tests of foraminiferans, shells of molluscs and debris from
both. Planktonic and benthonic foraminiferans are both abundant but are
mainly small possibly juvenile, tests. Similarly, the molluscs fraction
consists mainly of juvenile shells and fragmental material; few adult
individuals are present. Terrigenous debris comprises up to 20% of the sand
fraction. The major sediment components of the sand fraction suggests a lack
of indigenous production of sediment in the inter-reef channels. Primary
sediment components originate mainly on the inner shelf or in the open ocean
and are transported into the channels by tidal and wind driven currents.
Absence of reef-derived materials in the channels emphasizes the lack of
sediment transport outward from the reef margins. Reef-derived sediments are
limited to the immediate vicinity of the reefs, decreasing rapidly in
abundance within about 1km of the reef edge.

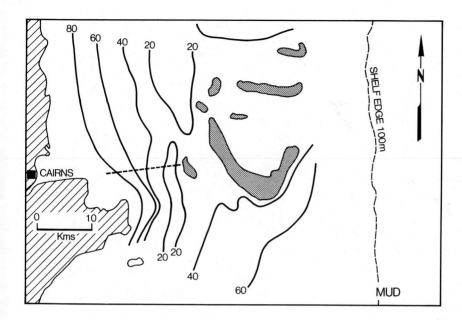

Fig. 3.   Percentage contribution of mud in shelf sediments (from Maxwell and Swinchatt, 1970).

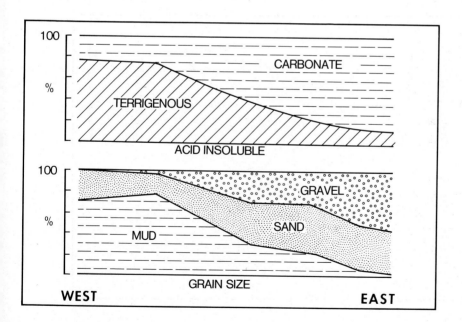

Fig. 4.   Variation of acid insoluble and grainsize of shelf sediments along one traverse (see Fig. 3 for location).

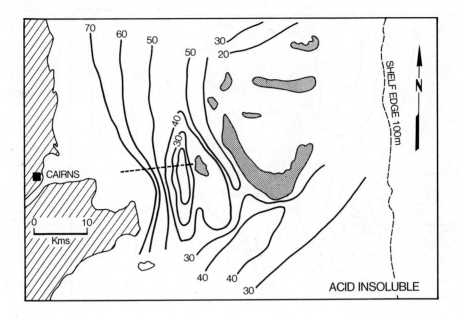

Fig. 5.   Percentage contribution of acid insolubles in shelf sediments
(from Maxwell and Swinchatt, 1970).

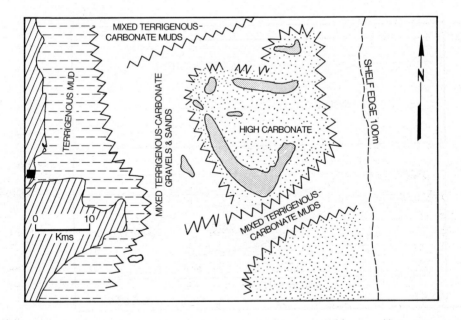

Fig. 6.   Distribution of lithofacies on the shelf near Cairns. The mixed
terrigenous-carbonate gravels and sands belong mainly to the
impure carbonate facies.

16 SHELF SEDIMENTS OF THE HOWICK GROUP

Details of the inter-reef sediments of the shelf in the vicinity of the
Howick Group (Fig. 7) have been described by Flood et al (1978). Even though
the across-the-shelf variation in the terrigenous and carbonate contribution
resembled that of Maxwell and Swinchatt (1970) for the Cairns region
(compare Figs. 4 and 8), it was apparent that the sediments consist of
material derived from at least three sources:

1. Terrigenous material (predominantly mud) derived from a variety
   of igneous and sedimentary rock types which occur on the mainland and continental islands.

2. Carbonate material representing the calcareous skeletons
   of in situ organisms such as Halimeda, molluscs, byrozoans,
   echinoderms, ostracods, solitary corals, benthis foraniniferans,
   etc. The foraminiferans Marginopora vertebralis and Alveolinella
   quoyi are ubiquitous (the latter be restricted to the shelf
   sediments).

3. The reef derived material includes a significant contribution
   from corals, molluscs, Halimeda, benthonic foraminiferans such
   as Calcarina hispida and Baculogypsina sphaerulata. Abundance
   of these foraminiferans provide a relatively precise indication
   of reefal proximity.

An across-the-shelf plot of the acid-insoluble content of the sediments
(Fig. 9) far removed from the influence of reefal contributions revealed
some interesting results. Firstly the terrigenous mud component extended
approximately two-thirds of the way across the shelf to abruptly stop within
a 2 km zone. The terrigenous gravel and sand component of the sediments
mirrored this pattern. Obviously across-the-shelf transport of terrigenous
sediment was not operating in this area. Secondly, within the gravel and
sand component of the sediments up to 60% was acid soluble material, mainly
the benthic foraminiferans, Marginopora and Alveolinella. These findings
showed for the first time that there existed a significant in situ
post-mortem contribution of skeletal carbonate to the shelf sediments.
Thirdly, detailed examination of the shelf sediments revealed that the
contribution of the reefs to the adjacent shelf sediments was restricted to
within a few km of the reef (Fig. 10).

The relatively uniform terrigenous and in situ carbonate contribution
throughout the area is overshadowed in the vicinity of individual reefs by a
series of concentric zones characterized by higher carbonate content,
increasingly coarser grain size and improved sorting (reflecting the
predominance of one source). The pattern can be further complicated in areas

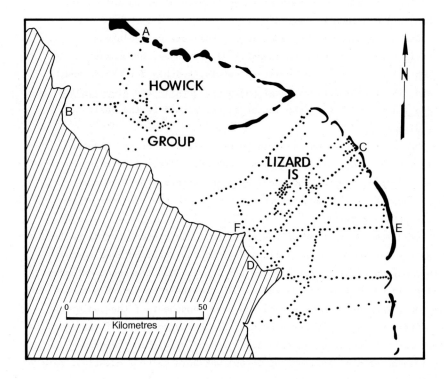

Fig. 7.   Pattern of sediment samples across the shelf in the area north of
Cooktown to Cape Melville. Letters A-F refer to transects showing
acid-insoluble content of shelf sediments.

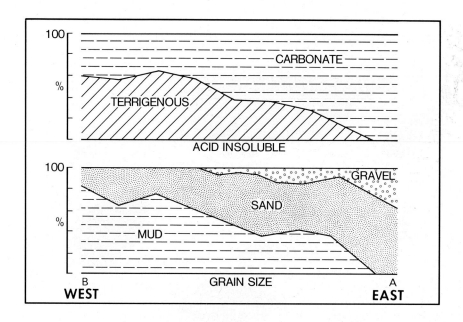

Fig. 8. Showing distribution of acid insoluble content and grain size of shelf sediments along traverse A-B, far removed from the influence of mid-shelf reef, Howick Group.

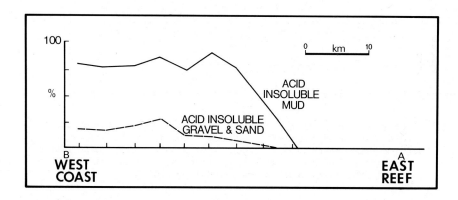

Fig. 9. Showing acid insoluble content of mu sizes and gravel and sand sizes along same traverse line as in Fig. 8.

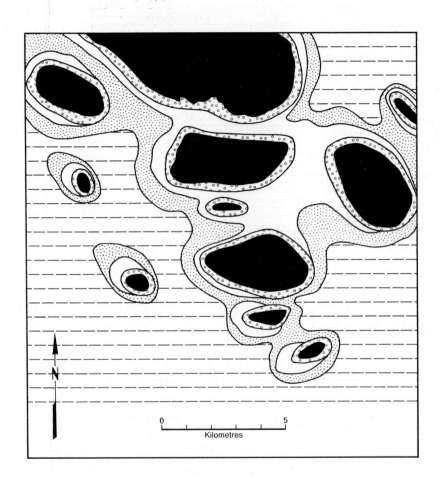

Fig. 10. Idealized distribution pattern of sediments around the mid-shelf
reefs. The influence of the reef mass (solid) on the shelf
sediments is restricted to about 2 km. Dashed lines, shelf facies;
open circles, proximal talus deposit; open, leeward sediment cone;
dots, distal talus deposit. Modified from Flood et al. (1978).

of dense reefal development where overlapping of the concentric arrangement occurs.

Surrounding each reef is a narrow (100m) zone consisting of coarse coral detritus to windward and reef-derived benthonic foraminiferans to leeward and represents a talus slope type of accumulation at the base of the reef slope. A characteristic sediment cone is developed to leeward. It consists of both traction and saltation transported reef flat material characteristically of sand sizes. Surrounding the entire reef in a concentric zone is a broader zone varying in width depending on the size of the reef mass, which contains the reef derived particle sizes which have been transported in suspension from the reef top during cyclones, ebb tidal flow, or severe storms. This material is characteristically fine sand and silt sizes. In general the concentric-elliptical patterns surrounds each reef in a zone approximately 2-3km wide in which there is a size gradient from coarse sand and gravel near the windward edge of the reef to fine sand near its leeward margin.

17 THE HALIMEDA BANKS

It was during the 1973 Royal Society of London and Universities Expedition to the northern reef that the occurrence of extensive banks of Halimeda debris (Fig. 11) were discovered towards the outer shelf in the lee of the outer barrier (ribbon) reefs. Halimeda debris is ubiquitous. The acme of production of this skeletal component occurs on a series of submarine banks. These banks occur at a fairly uniform depth of 20m, with intervening troughs and hollows descending to -37m.

The morphology of these banks is best observed on high-resolution seismic profiles (Fig. 12). They are up to 6km wide and extend as a series of ridges and troughs for about 100km behind the shelf-edge reefs. Initially it was considered that they could represent calcareous dunes developed during the Pleistocene low stands of sea levels, but recent vibrocoring results have clearly demonstrated that the entire mass consists of unlithified Halimeda flakes. These banks represent some 15m of vertical accumulation of carbonate sediments during the past 10,000 years when the sea flooded the shelf during the Holocene transgression.

The amount of $CaCO_3$ produced by these banks is several orders of magnitude greater than that of the shelf edge reefs which are usually less than 1km wide. Recent calculations (Drew and Abel, 1983) of the rate of productivity of reefal Halimeda of $2.5 \pm 1.1 kg\ m^{-2} yr^{-1}$ is almost the $4.0 \pm 0.3 kg\ m^{-2}$ produced by the entire coral reef system (Davies and Marshall, 1979).

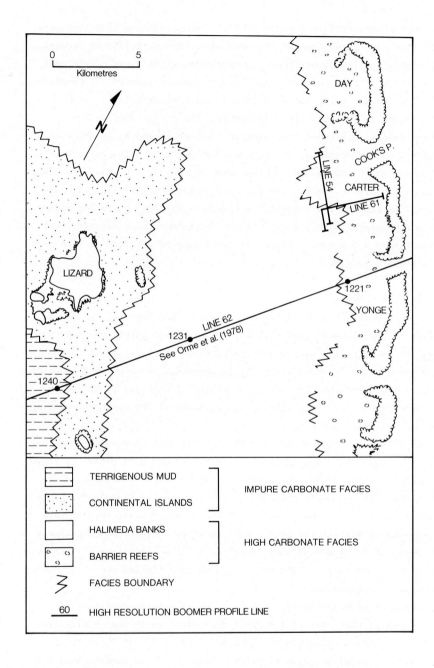

Fig. 11.   Distribution of lithofacies between Lizard Island and the outer
           barrier reefs. Line 62 is illustrated in Orme et al. (1978).

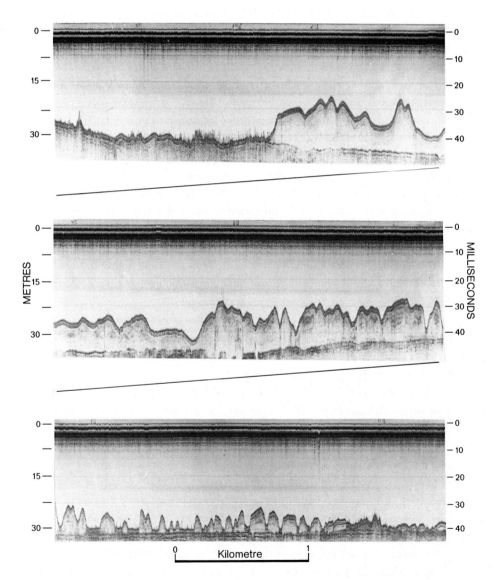

Fig. 12. High resolution seismic profile across part of a Halimeda bank from near the outer reef (top left) to mid-way across the shelf (lower right).

Clearly these Halimeda banks have an enormous influence on the production of carbonate sediments, particularly mud-size sediment, and could overshadow the influence of the shelf edge reefs as a source of carbonate to the shelf sediments.

Orme et al (1978) have examined the distribution of the shelf facies in the area extending from Lizard Island to the shelf edge. On the basis of the relative proportions of carbonate and terrigenous end-members present they recognize a primary distinction between high carbonate and impure carbonate facies in the manner of Maxwell (1968).

The high carbonate facies includes two subfacies, namely the Halimeda gravels and sands of the submarine banks and the coral-algal gravels and sands of the back-reef environment of the shelf edge. The former sub-facies is dominated by whole and fragmented Halimeda plates, some gastropods, pelecypods, numerous and varied foraminiferans together with minor components including bryozoans, echinoid spines and tests, spicules and a few quartz grains. These sediments are very well sorted and unimodal, owing to the dominance of the Halimeda debris, which points to the importance of ecological factors favouring the luxuriant growth of Halimeda on these self-perpetuating banks. The coral-algae subfacies is dominated by algae coated coral sticks, Halimeda, molluscs, and reefal benthonic foraminiferans. These sediments represent the complex interplay of the current of delivery, which transports sediment from the reef tops, and the current of removal, which winnows out the finer grained sediment to deposit it further from the reef in the back-reef environment.

Sediments of impure carbonate facies in this area are extremely varied owing to the varied contribution of sediment from several sources, namely, Halimeda banks, fringing coral reefs, continental islands, mainland terrigenous mud, and in situ organisms. Near the continental islands some sediments display proportions of acid insolubles; mainly quartz, that would place them in the transitional or terrigenous facies but the distribution of such sediments is very localized. For example, immediately to the northwest (leeward side) of Lizard Island, away from the influence of fringing reefs, the sediments are dominated by terrigenous sand derived from the granitic rocks of Lizard, Palfrey, and South Islands. Quartz of fine to medium sand grade dominates this facies, but feldspar, tourmaline, and rock fragments also occur. The carbonate fraction is composed largely of molluscs and foraminifera. Farther west and southwest from Lizard Island the sediments are bimodal containing a high proportion (often exceeding 50%) of terrigenous mud derived from the mainland and a coarse carbonate fraction composed of mollusc, foraminiferans, and Halimeda debris.

18 GENERAL SHELF FACIES PATTERN

The general nature of the sedimentary facies on the continental shelf may
be established by examining in excess of 250 sediments collected during the
1973 Expedition. These sediments were collected systematically using a
Peterson Grab and sample spacings of approximately 1.5km across the shelf on
several traverse lines (see Fig. 7). In several instances the traverses
coincided with seismic (high resolution boomer) profiles. More recently,
during 1981, approximately 50 shallow (4m) vibrocores were obtained across
this part of the shelf. These cores are being studied at the present time.

The sediments were treated with hydrogen peroxide to remove carbonaceous
matter in preparation for sieve analyses. The samples were wet sieved and
the percentage gravel, sand and mud calculated. In addition, textural
statistics were compiled from grain size data using $0.25\emptyset$ U.S. Standard
Sieves. The samples were split and treated with dilute hydrochloric acid to
remove any calcium carbonate present. Percentage composition of terrigenous
(acid insoluble) and carbonate components were calculated.

Selected samples were scrutinized microscopically in order to ascertain
whether preferential size grades were adopted by certain grain types, which
might in turn relate to ecological and/or dispersal factors.

The acid insoluble content of the gravel plus sand fraction and the mud
fraction of the shelf floor sediments display a well marked drop off of the
terrigenous influence approximately mid way across the shelf. On the outer
half of the shelf any departure from the norm can be explained by the
localized influence of the high continental granitic islands such as Lizard
and South Direction Islands (Fig. 13). This marked decrease in the acid
insoluble mud content of the sediments dispels the earlier suggestions of
across-the-shelf sediment transport. The same conclusion was reached by
Belperio (1983) who was describing dominance of the vertical and lateral
progradation of coastal sediments over shelf sedimentation in the Great
Barrier Reef lagoon.

A generalized sediment facies distribution pattern for this area of the
northern Great Barrier Reef is presented in Fig. 14 and selected sediments
are illustrated in Figs. 15-16.

On the inner-shelf and near-shore the Terrigenous Facies is dominated by
wind blown quartz sand and reworked fluvial quartz sand. Sediments of this
facies are moderately well sorted coarse to fine sands with minor gravel and
mud sized components. They extend seaward as a prograding wedge to
approximately the 10m bathymetric contour. At this point the sea bed
increases rapidly and only the finest sand and silt sized quartz particles
extend beyond the break in slope.

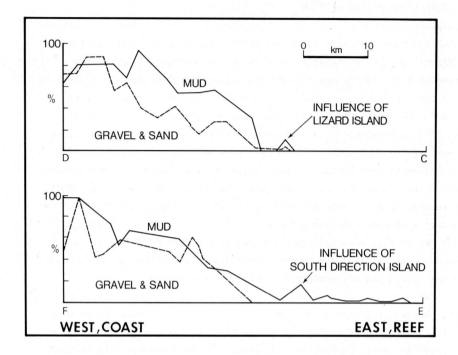

Fig. 13. Across the shelf traverses showing the abrupt break in the acid insoluble content of the shelf sediments. The influence of the continental islands is obvious.

The Transitional Facies displays variable contributions from both the mainland derived mud and in situ calcareous skeletal organisms. The gradational contact of this facies and the Impure Carbonate Facies represents the seaward limit of transport of sand sized quartz particles.

The Impure Carbonate Facies is an extremely heterogeneous admixture of sediment derived from a variety of sources including terrigenous mud, in situ post-mortem skeletal contribution, distal to proximal reefal influence, continental islands, and carbonate mud derived from the Halimeda banks. In restricted areas it is possible to recognize distinctive subfacies such as reefal carbonate, quartz sand, terrigenous mud, carbonate mud, etc.

The High Carbonate Facies is restricted to the outer half of the continental shelf. Two distinctive subfacies can be recognized. Firstly, the Halimeda gravel and sand subfacies which corresponds with the areal extent of the Halimeda banks, and secondly, the reefal skeletal carbonate sediment which changes from mud to sand to gravel as the reef complex is approached.

199

In some areas near the passages between the shelf edge reefs the sea
floor is a current-swept rugged surface of recrystallized Pleistocene reef
mass.

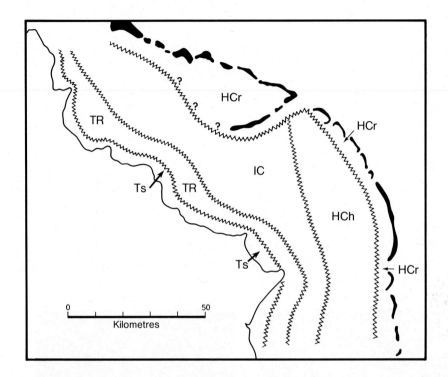

Fig. 14. Distribution pattern of the lithofacies on this portion of the
shelf. Ts, terrigenous (sand); TR, transitional; IC, impure carbonate, HC,
high carbonate, r, reefal and h, Halimeda subfacies.

19 CONCLUSIONS

1. Lithofacies maps based merely on the proportions of carbonate and
siliciclastic components present an oversimplified picture of an
across-the-shelf transition from mainland terrigenous derived siliciclastics
to outer-shelf edge derived carbonates.

2. The decrease in the contribution of the siliciclastic components to
the shelf sediments is not gradual. This is a consequence of a variety of
factors principally the dominance of shore-parallel sediment transport of
the quartz sands in the near-shore zone and terrigenous muds on the
inner-shelf.

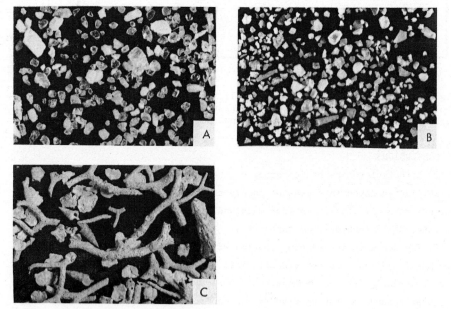

Fig. 15.   Photographs of sediments (mud removed) selected to illustrate the
range of lithofacies. A, terrigenous sand; B,C, transitional. Fields
of view respectively are A, 6 mm; B,C, 80 mm. The contribution of
in situ foraminiferal material is obvious in samples B and C.

Fig. 15(cont). Photographs of sediments (mud removed) selected to illustrate the range of lithofacies. D, E, impure carbonate; F, impure carbonate displaying contribution of reef top skeletal debris. Fields of view respectively are D,E, 80 mm; F, 16 mm. The contribution of _in situ_ foraminiferal material is obvious in samples D and E.

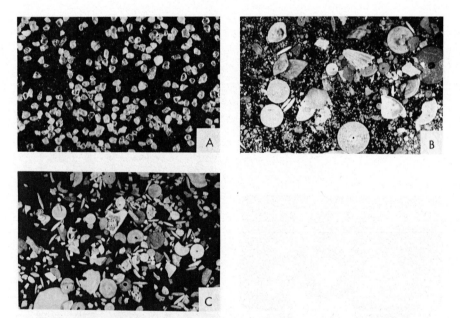

Fig. 16.   Photographs of sediments (mud removed) selected to illustrate the
range of lithofacies. A, impure carbonate with quartz sand in
vicinity of a high continental island; B, impure carbonate in
vicinity of a fringing reef on a continental island; C, high
carbonate in a trough within the <u>Halimeda</u> banks, gastropods and
bryozoans are evident. Fields of view respectively are A, 6 mm; B, 16
mm; C, 80 mm.

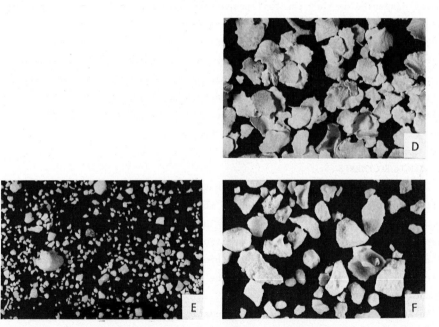

Fig. 16(cont). Photographs of sediments (mud removed) selected to illustrate
the range of lithofacies. D, high carbonate of the <u>Halimeda</u> banks; E, high
carbonate, distal sediment apron of outer barrier reef; F, high carbonate,
proximal sediment apron of back reef. Fields of view respectively are D, 80 mm;
E, 2 mm; F, 16 mm.

204

3. The carbonate component reflects not only the reefal contribution but
also the _in situ_ post-mortem contribution of molluscs, bryozoans and
profilic benthonic foraminiferans, particularly the genera _Marginopora_ and
Alveonella. On the outer half of the shelf there is a significant
contribution of gravel and sand sized particles to the _in situ_ accumulation
of _Halimeda_ flakes - whilst the finer mud size particles may be transported
some distance to be mixed with terrigenous muds in the area of the Impure
Carbonate Facies.

4. There is a minor but locally significant contribution of siliciclastic
material made by the continental high islands to the mid-shelf sediments.

5. Mixing of end-member sediment types is an important process in
determining the nature of most inter-reef sediments. Some sediment
accumulations may represent lag deposits (e.g. _Halimeda_ gravels), others are
strongly dependent upon post-mortem contributions (e.g. inter-reef benthonic
foraminiferans), and some are the product of currents of delivery and
removal (e.g. the winnowed back-reef sediment aprons).

6. The degree of dispersal of sand-sized sediments from the outer barrier
reefs, the mid-shelf reefs, and the high continental islands is limited to a
few kilometres from their source. The finer-grained particles would be more
widely dispersed as they can be transported considerable distances (tens of
kilometres) in suspension by the tidal currents.

7. The spatial arrangement of the sedimentary facies described here
represents but one instant in time. Work in progress utilizing the results
of the vibrocores through the shelf sediment which vary from one to several
metres thick will enable the temporal pattern of facies changes to be
established. Such a study will provide extremely valuable data on the
clastic to carbonate to clastic facies changes.

20 ACKNOWLEDGEMENTS

The major findings of this research have resulted from the author's
participation in the 1973 Expedition of the Royal Society and Universities
of Queensland to the northern Great Barrier Reef. Since 1979, extra high
resolution seismic investigations, additional sediment sampling, and shallow
vibrocoring of the shelf sediments has been financed by grants from the
Australian Research Grants Scheme and the Australian Marine Science
Technology Advisory Committee.

An Australian-American Education Foundation Fulbright Grant in 1981
enabled PGF to inspect modern and ancient reef in the Caribbean thus
allowing comparisons with the Great Barrier Reef.

REFERENCES

Belperio, A.P., 1983. Late Quaternary terrigenous sedimentation in the Great
    Barrier Reef Lagoon. In: Baker, J.T., Carter, R.M., Sammarco, P.W. and
    Stark, K.P., (Editors), Proceedings: Inaugural Great Barrier Reef
    Conference, Townsville, JCU Press, pp.71-76.
Bird, E.E., 1971. The fringing reefs near Yule Point, north Queensland.
    Aust. Geog. Stud., 9, pp. 107-115.
Davies, P.J. and Martin, K., 1976. Radial aragonite ooids, Lizard Island,
    Great Barrier Reef, Queensland, Australia. Geology, V. 4, pp. 120-122.
Drew, E.A. and Abel, K.M., 1983. Growth of Halimeda in reefal and
    inter-reefal environments. In: Baker, J.T., Carter, R.M., Sammarco, P.W.
    and Stark, K.P., (Editors), Proceedings: Inaugural Great Barrier Reef
    Conference, Townsville, JCU Press, pp. 299-304.
Flood, P.G., 1983. Coated grains from the Great Barrier Reef. In: Peryt,
    T.M. (Editor), Coated Grains. Springer-Verlag, Berlin, pp. 561-565.
Flood, P.G., Orme, G.R. and Scoffin, T.P., 1978. An analysis of the textural
    variability displayed by inter-reef sediments of the impure carbonate
    facies in the vicinity of the Howick Group. Phil. Trans. R. Soc. Lond.
    A., 291, pp. 73-83.
Ginsburg, R.N. and James, N.P., 1974. Holocene Carbonate Sediments of
    Continental Shelves. In: Burk, C.A. and Drake, C.L., (Editors), The
    Geology of Continental Margins, Springer-Verlag, Berlin, pp. 137-155.
Maxwell, W.G.H., 1968. Atlas of the Great Barrier Reef. Elsevier.
Maxwell, W.G.H., 1973. Sediments of the Great Barrier Reef. In: Jones, D.A.
    and Endean, R., (Editors), Biology and Geology of Coral Reefs, V. 1,
    Geology 1, Academic Press, New York, pp. 299-345.
Maxwell, W.G.H. and Swinchatt, J.P., 19 0. Great Barrier Reef; regional
    variation in a terrigenous-carbonate province. Bull. Geol. Soc. Am., V.
    81, pp. 691-724.
Orme, G.R., Flood, P.G. and Sargent, G.E.C., 1978. Sedimentation trends in
    the lee of outer (ribbon) reefs, Northern Region of the Great Barrier
    Reef. Phil. Trans. Roy. Soc. Lond. A, V. 291, pp. 85-99.
Orme, G.R. and Flood, P.G., 1980. Sedimentation in the Great Barrier Reef
    Province, Adjacent Bays and Estuaries, In: Henderson, R.A. and
    Stephenson, P.J., (Editors), The Geology and Geophysics of Northeastern
    Australia. Geol. Soc. Aust. Qd. Div., Brisbane, pp. 419-434.
Stoddard, D.R., 1965. British Honduras cays and the low wooded island
    problem. Trans. Inst. Brit. Geogr., V. 36, pp. 131-147.

Chapter 7

INFILLING OF COASTAL LAGOONS BY TERRIGENOUS SILICICLASTIC AND MARINE
CARBONATE SEDIMENTS: VIEQUES, PUERTO RICO

G.M. D'ALUISIO-GUERRIERI and R.A. DAVIS, JR.
Total Petroleum, 2950 1 Allen Center, Houston, Texas 77002
Department of Geology, University of South Florida, Tampa, Florida 33620

ABSTRACT
    Coastal lagoons on the southern coast of Vieques, Puerto Rico, present an
excellent opportunity to study the relative contributions of terrigenous
sediments and carbonate sediments to the infilling of these tropical
lagoons. The four lagoons studied are formed by ridges of Eocene limestones
in combination with tombolos and other Holocene sediment accumulations. Each
has a narrow opening to the open marine environment, which restricts tidal
circulation.
    Initial sediment provided to the lagoon is in the form of terrigenous mud
and sandy mud derived from runoff from the highlands to the north. As sea
level invaded the lagoons, carbonates began to accumulate, some generated in
the lagoons as marine biogenic debris and some from the adjacent fringing
reef complex in the open marine environment. Storm tidal deltas developed
landward of the constricted openings between the lagoons and the open sea.
As shallow, turbid waters gave way to clear water, sea grass (Halodule) was
replaced by Halimeda in the central and seaward parts of the lagoons. Storm
deposits are interbedded with sea grass peats and Halimeda sand.
    Mangroves became abundant along the intertidal fringes of the lagoons at
least several hundred years ago. A recent increase in terrigenous runoff has
caused the mangroves to be stressed and terrigenous siliciclastic sediments
to prograde over lagoonal carbonates.

1 INTRODUCTION

    Sedimentation along coasts, especially embayed coasts, is a result of

complex interactions of processes and sediments. Commonly, coastal

embayments represent geologically ephemeral environments that are filled by

a combination of land-derived and marine-derived sediments. Detailed study

of the modern sediment facies combined with examination of the stratigraphy

of the sediments that have accumulated in coastal embayments, permits

reconstruction of the process-response systems that operate in such a

complex coastal system.

    Estuaries typically receive significant and continuous runoff from the

land and experience good tidal circulation with the open ocean. Lagoons, on

the other hand, do not receive regular runoff from land, and typically have

a restricted circulation with the open marine environment. The southern

coast of Vieques, Puerto Rico, contains numerous small lagoons which afford

an excellent opportunity to examine the sedimentologic and stratigraphic

record of sediment infilling in these coastal embayments. Because the marine

system is dominated by biogenic carbonate sediment and the terrigenous
source can be easily recognized, these lagoons are ideal for the study of
relative marine and terrigenous sources. As such these lagoons provide an
excellent example of local transition between terrigenous and carbonate
sediments.

2 OBJECTIVES

Four lagoons along the southern shore of the island of Vieques, Puerto
Rico were chosen for study. They were selected because they display varying
degrees of infilling by marine carbonate sediments and by terrigenous
alluvial sediments. The study lagoons also show recent modification by
formation of tombolos, spits, and beach ridges.

The fact that these environments are presently changing makes them an
excellent source of information for documenting, in detail, the geologic
development of lagoons on Vieques, primarily through stratigraphic and
sedimentologic study. This study establishes trends in lagoonal infilling
and reasons for past and possible future changes in the coastal lagoons of
Vieques.

3 LOCATION AND GENERAL DESCRIPTION

Vieques, the largest subsidiary island of the United States Commonwealth
of Puerto Rico, is located 10 km east of the main island. It is 19 km long
from east to west and 5 km wide at the middle (Fig. 1). The eastern half of
Vieques is covered with low thorn scrub of Zizyphus, Prosopis and Acacia
(Woodbury, 1972) with a more tropically vegetated forest to the west. This
distribution is the result of the high terrain in the west, 300 m above sea
level, where rainfall is approximately 30 percent greater per year than in
the east (Benitez, 1976).

With the exception of the eastern third, the northern coast of Vieques is
bordered by steep, poorly developed, narrow beaches. The eastern part of the
north coast is similar to the southeastern coast; it is composed of rocky
headlands, arcuate embayments, and beach ridges. The extreme northeast and
southeast shorelines display sea cliffs up to 50 m high, whereas the western
side is predominantly low beach and mangrove swamp.

The embayments of the southern shoreline receive sediments from headland
erosion and subsequent transport by longshore currents. The northern
shorelines, on the other hand, are largely erosional, with poorly developed
beaches. Vieques experiences a mean tidal range of 0.24 m. Tides are not a
major factor in controlling present beach or lagoon morphology.

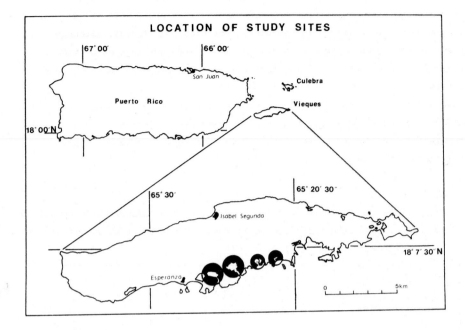

Fig. 1. Location of Vieques showing study sites (black circles) on the
south side of the island.

Marine carbonate sediments are comprised primarily of skeletal sand and
gravel. Adjacent to and in the mouth of lagoons, carbonate gravel is formed
by coral and molluscan debris from nearshore marine environments. Within the
protected lagoonal environment this carbonate gravel facies grades into
Halimeda sand and carbonate mud.

Terrigenous alluvial sand and mud are found along the landward margins of
lagoons. This sediment is best developed in the alluvial flats surrounding
the lagoons. Where terrigenous sediment has breached fringing populations of
mangrove, a facies of pelleted mud has developed. This facies grades into
carbonate mud, a transitional facies between terrigenous and
marine-dominated environment.

4 GENERAL GEOLOGY

The eastern half of Vieques is composed of Upper Cretaceous-Eocene
volcanic and pyroclastic rocks, with a large granodiorite body in the
central area. The western half is almost totally granodiorite and tonalite
and is covered in many areas by Quaternary alluvium (Briggs, 1964).
Oligocene-Miocene limestone forms the sea cliffs of the easternmost part of

Vieques and creates low-angle cuesta-forms along the central third of the southern shore (Fig. 1). Limestone commonly forms islands, some of which are attached to the main island by tombolos, thus creating lagoons. Although the serpentine complexes described by Briggs (1964) were not located on Vieques during field work, pebbles of serpentine were found imbedded in beach rock deposits.

Beach rock is present along 10 percent of the shoreline of Vieques and is commonly composed of <u>Halimeda</u> and shell fragments. Other coastal accumulations are dominated by clasts of various rock fragments, biogenic debris and quartz sand.

5 STUDY SITES

Although four lagoons were studied, only two will be discussed in detail in this report. The westernmost lagoon is Puerto Mosquito (Fig. 2). It is

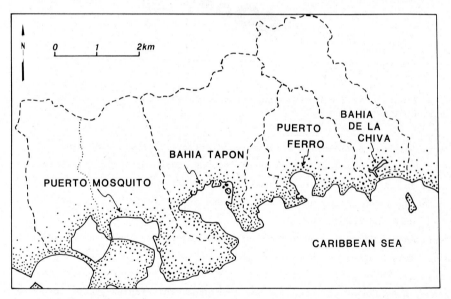

Fig. 2. General geography of study lagoons with drainage basins for each shown by the dashed line.

bounded on the southwest and southeast by Oligocene-Miocene limestone which forms cuesta-like structures that dip gently to the south. The western section of Puerto Mosquito received open marine circulation until it was cut off by a tombolo. It now shares limited circulation with the main body of Puerto Mosquito through a narrow mangrove filled channel (Fig. 3).

To the north of Puerto Mosquito are tonalite and granodiorite bodies which form hills 100 to 150 m high. Parts of these hills are covered by Quaternary colluvium and alluvium. The alluvium extends to the mangrove fringe bordering Puerto Mosquito, and forms a barren zone that contains

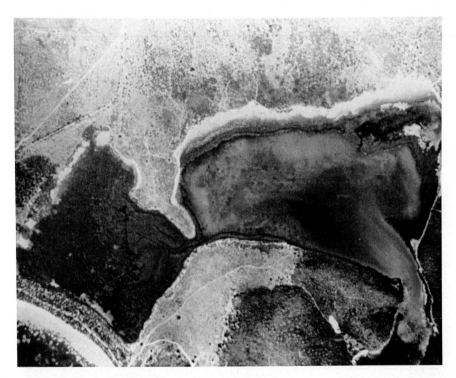

Fig. 3. Vertical aerial photo of Puerto Mosquito.

hypersaline pools impounded landward of the mangroves. Puerto Mosquito is linked to the Caribbean by a wide, shallow (1 to 2 m) inlet. This inlet breaches the limestone cuestas that form the southern border.

Puerto Ferro is east of Puerto Mosquito (Fig. 2). It is bounded on the southeast by the same cuesta-like limestone body that borders the southeast of Puerto Mosquito. To the east, a similar limestone body is also present, but it is lower in profile and slope. The northern terrain is composed of the same tonalite and granodiorite bodies covered by Quaternary colluvium and alluvium. The alluvium extends south across the alluvial flats of Puerto Ferro and ends at the mangrove fringe. Hypersaline pools similar to those in

212

Puerto Mosquito are impounded behind the mangrove fringe of Puerto Ferro.
Puerto Ferro receives circulation through a wide channel that is 2 to 4 m
deep (Fig. 2).

Bahia Tapon is separated from Puerto Ferro by a wide embayment (Fig. 2).
The western boundary is a low, heavily weathered, Oligocene-Miocene
limestone surface. Bahia Tapon is bordered on the north and east by hills of
upper Cretaceous-Eocene volcanic and pyroclastic rocks and by outcrops of
limestone (Fairbridge, 1975). The barren alluvial flat on the north of the
Bahia Tapon mangrove fringe is narrower than those flats already described,
but to the east, it expands into a wide, shallow, hypersaline pool (Fig. 4).

Fig. 4. Oblique aerial photo of Bahia Tapon looking to the northwest.

A small island, colonized by mangroves, is present in the center of the
lagoon. Bahia Tapon receives circulation through a narrow channel that is 20
to 30 m wide and 1 m deep at its center.

Bahia de la Chiva is the easternmost lagoon studied (Fig. 2). Three minor
headlands and two small, nameless embayments separate this lagoon from Bahia
Tapon. Bahia de la Chiva is bounded on the east and west by Upper
Cretaceous-Eocene lavas and pyroclastics. A belt of Quaternary alluvium
extends from the north along a wide valley into this narrow lagoon. Bahia de
la Chiva is surrounded by a narrow fringe of red mangrove, Rhizophora
mangle, which is in turn surrounded by a wider band of black mangrove,
Avicennia germanin. To the south of Bahia de la Chiva, a series of beach
ridges extends parallel to the lagoon and separates it from the Caribbean. A
narrow inlet, 25 m wide, existed between the western end of the lagoon and
the Caribbean until a road was constructed across the channel in the 1940's.
Despite a conduit beneath the road, Bahia de la Chiva has contact with the
Caribbean Sea only when storm runoff breaches both the road and the
confining beach ridges.

6 DATA COLLECTION AND ANALYSIS

Field work was conducted during the summer of 1981. Emphasis was on
mapping the bathymetry and surface sediment facies of the lagoons in order
to characterize present conditions. Cores were also taken across each of the
lagoons in order to construct a stratigraphic framework for interpreting the
history of infilling of the lagoons.

6.1 Surface Samples and Bathymetry

Sediment samples and bathymetric data were collected in order to obtain
information for construction of a map of modern sediment facies and to
consider any controls on sedimentation that may be affected by bottom
topography. Samples were collected from the upper 3 cm of the sediment.

Bathymetry was measured using a calibrated rod in water less than 1.5 m
deep in depth and a weighted nylon tape measure for depths greater than 1.5
m. These measurements were based on a datum of mean low tide with the aid of
tide charts and from a time-frequency graph constructed with data from a
reference gauge placed in the lagoon being sampled.

6.2 Coring

Cores provided the bulk of the information for the study of lagoonal
history. The coring rig included a sheet of plywood, an aluminum tripod with
telescoping lens, a rubber piston and a 6.1 m steel rod attached to a 6.1 m
length of chain, a 1800 kgm capacity come-along, a long handled shackle and
two 2 m lengths of chain. The coring tube was 7.62 cm (3 in.) diameter
aluminum irrigation pipe.

Cores were taken in transects perpendicular to the strike of modern depositional environments, starting in the northern alluvial flats and terminating in the open lagoon.

6.3 Bedrock Probing

A depth-to-bedrock survey was conducted in the four study lagoons to determine if bedrock exerts any control on lagoon morphology. Although the Oligocene-Miocene limestone and Upper Cretaceous-Eocene volcanic rocks define all but the northern borders of the four lagoons and the southern border of Bahia de la Chiva, questions remained about possible controls exerted on the northern borders and open lagoon bathymetry.

Probing/penetration ranged from 3 to 5 m in the open lagoons and less along the alluvial flats that border them. Bedrock was never reached. Gravelly colluvium was encountered in several areas but a clayey sand proved an impenetrable barrier in most lagoons. This barrier also limited the depth to which the coring tube could be driven.

7 CORE ANALYSIS

The first step taken in logging cores was to correct for compaction. This correction varied greatly, depending on lithology. Rhizophora peat compacted 60 to 70 percent whereas the chemically reduced, sandy clay facies showed essentially no compaction. A compaction index was created for each facies from information gathered in the field.

Once a general facies compaction index had been determined, the apparent thickness of each facies within a core could be multiplied by the appropriate index. The sum of all these facies index products invariably came within about 0.1 m of the actual penetration of the core. Because these calculated values better represent true thicknesses than treating compaction as a uniform phenomenon, they were used to construct a scale for logging.

Faunal and floral assemblages were identified by macroscopic examination in an attempt to find species that are abundant yet sufficiently restricted to delineate facies. Faunal remains were assumed to be close to the location of expiration in the low-energy environments due to the presence of many articulated pelecypods. This assumption is supported by Warme (1969), who showed that post-mortem transport in lagoonal environments is insignificant for most paleontological studies. A variety of shallow marine molluscs was identified. Macrofauna other than molluscs were rare with the exception of coral, which is abundant in storm deposits. Microfauna were not investigated.

8 BATHYMETRY

8.1 Puerto Mosquito

The bathymetry of the main body of Puerto Mosquito includes extensive shallow areas. These areas are located in the north and northwest zones of the lagoon and are less than 0.5 m deep (Fig. 5). The southeastern part of

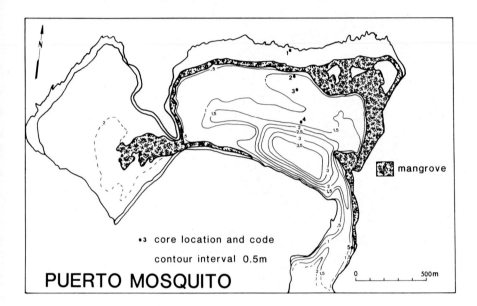

Fig. 5. Bathymetry and core locations, Puerto Mosquito. Compare with Figure 3.

the main body is nearly 4.0 m deep. This appears to be a remnant of an earlier, deeper lagoon. It remains sediment starved due to resistant limestone bedrock to the south and the relatively great distance from alluvial sediment sources to the north and northwest. It may also represent the remnant of a tidal channel created by the prism of a once deeper western part of Puerto Mosquito. The inlet of Puerto Mosquito is shallow and is filled with sediments from adjacent offshore and lagoon environments.

The depth is 0.5 m where the perimeter of the main body of Puerto Mosquito is defined by mangrove roots. Western Puerto Mosquito is brackish and is less than 0.3 m deep.

216

8.2 Bahia Tapon

The bathymetry of Bahia Tapon is controlled, at least in part, by storm deposition. In the southern half of the lagoon, this storm generated accumulation approximates the morphology of a flood delta (Fig. 6). The

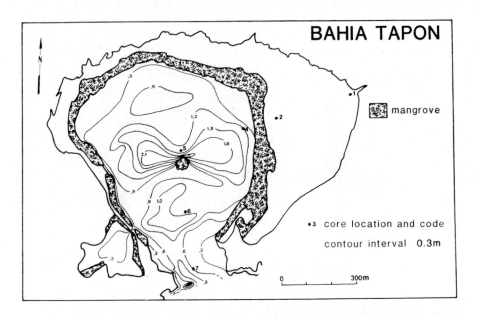

Fig. 6. Bathymetry and core locations, Bahia Tapon. Compare with Figure 4.

shielding effect of this flood delta is seen in the bathymetry directly behind the central island of the lagoon. This island is comprised of storm-derived limestone and debris from adjacent environments. Boulder deposition, also the result of storm activity, is present on the west side of the inlet at Bahia Tapon. Where mangrove roots define the borders of this lagoon, the depth of the perimeter is 0.5 m or greater (Fig. 6).

The northern and eastern sections of Bahia Tapon contain broad, shallow subtidal to supratidal flats. A narrow, east-west trending area of about 2 m depth represents the remnants of the deeper lagoon.

8.3 Puerto Ferro and Bahia de la Chiva

Although Puerto Ferro and Bahia de la Chiva were studied in the same detail as Puerto Mosquito and Bahia Tapon, they will not be discussed extensively in this report. Puerto Ferro is the deepest of the four lagoons (4.5 m) and has the greatest circulation with the open marine system. The

regular bathymetry approximates that of an oval basin. Bahia de la Chiva is
the smallest, shallowest, and has the most restricted circulation of the
four study lagoons. The main body of this lagoon is less than 0.5 m deep;
however, the connecting inlet exceeds 1.5 m in depth.

9 LAGOONAL FACIES

Sedimentary facies are delineated by grouping sediment samples that
exhibit similar grain-size distribution and organic and mineral content. All
facies are not present in all four lagoons. More than 200 surface samples
were analyzed in determining sediment distribution patterns.

The stratigraphic units described in core logs have been grouped into ten
facies similarly based on biota and sediments. Components used in defining
facies are those that commonly occur in association and display good
correlation between cores. (Figs. 7 and 8). This terrigenous mud contains
some ostracods and algal mats. In areas proximal to peat and mangroves, it
becomes black, indicating reducing conditions and increasing organic
content. Terrigenous mud in all of the study lagoons commonly contains
traces of quartz sand and root material.

A zone of muddy sand is immediately landward of the terrigenous mud and
extends over the alluvial flats of Puerto Mosquito and Bahia Tapon (Figs. 7
and 8). This facies commonly decreases 0.2 to 0.3 m in elevation between the
scrub brush bordering the alluvial flats and terrigenous mud. Terrigenous
muddy sand closely parallels and often defines the boundary of hypersaline
pools. The sand fraction ranges from 30 to 90 percent in this facies and is
comprised of fragments of limestone, granodiorite, and grains of quartz and
epidote.

The subsurface terrigenous alluvium facies has been subdivided into two
groups: oxidized terrigenous alluvium and chemically reduced terrigenous
alluvium. Oxidized alluvium at or near the surface is light brown to medium
brown. Alluvium at depths of 1 to 2 m near the perimeter of the alluvial
flats is reduced displaying a bright blue-green color. Where alluvium occurs
beneath marine facies it also shows reducing conditions to a depth of 0.5 to
1.0 m. Oxidized alluvium was encountered below this reducing zone in Bahia
Tapon.

Terrigenous alluvium is equivalent in composition to the terrigenous
muddy sand, and all sediments were similarly derived from surrounding hills.
Alluvium is primarily a muddy sand within oxidized sediments below alluvial
flats, and a sandy mud in the reduced parts of this facies. Oxidized
alluvium found below reduced alluvium in Bahia Tapon is a red-brown, muddy,
sandy gravel.

The only macroscopic organic component found in terrigenous alluvium is scattered, unidentified terrestrial root material, which closely resembles that found in terrigenous sediments of the alluvial flats.

9.1 Mangrove Peat

Red mangrove (Rhizophora mangle) peat is present in all four lagoons studied. The peat is reddish-brown to reddish-black and is comprised of rootlets, leaf litter, and occasional prop roots. This composition closely matches descriptions by Davis (1940). The surface expression of mangrove peat is restricted to the base of mangroves and the landward perimeter of fringing populations where mangroves have recently expired. These surrounding fringes are 5 to 75 m wide. Extraneous material is uncommon within the peat, but scattered ostracod and crab shells are present. Terrigenous sediment causes landward parts of the peat to be muddy. The mangrove peat facies is generally 0.1 to 0.4 m higher in elevation than adjacent terrestrial environments.

The subsurface character of mangrove peat is identical in composition to the surface facies. Mangrove peat is red-brown in fresh samples and reddish-black when oxidized. Most of this unit contains no inorganic constituents. In certain zones, peat becomes muddy with terrigenous silt and clay. Anomalocardia brasiliana is common in these muddy peat zones. Balanus Eberneus is found in mud free zones. Rhizophora peat is indicative of past shoreline position and tidal range because it accumulates only within microtidal ranges (Scholl, 1964, b).

9.2 Terrigenous Alluvium

Terrigenous alluvium borders the study lagoons on all landward margins. The best developed examples of this facies are found in the alluvial flats behind fringing mangroves. In these areas, terrigenous alluvium can be divided into two distinct subfacies: terrigenous muddy sand and terrigenous mud.

Terrigenous mud is a facies of silt and clay present within the alluvial flats directly behind the mangrove fringe of Puerto Mosquito and Bahia Tapon.

9.3 Pelleted Mud

Pelleted mud is a transition facies between the terrigenous sediment impounded behind fringing populations of mangroves and the adjacent carbonate mud facies derived primarily from marine biogenic debris.

Terrigenous siliciclastic sediments breach the mangrove fringe during
periods of high runoff, depositing fines, which flocculate near the low
energy perimeter of the lagoons.

The mud fraction contains about 50 percent noncarbonate, terrigenous
sediments. The sand fraction ranges from 5 to 30 and is comprised of fecal
pellets, scattered quartz, and mollusc shell material. Pellets are cohesive
and are mostly carbonate. Pelleted mud is slightly sandy to sandy mud in
texture. No living benthic macro-vegetation was found in this environment.
The aqueous environment over this facies is low energy, but the water
exhibits high turbidity, primarily from the activity of benthic organisms.

The subsurface facies of pelleted mud is found directly below and is
identical to the pelleted mud sediments on the surface of the modern lagoon.
The stratigraphic position of this facies and the fact that it contains
about 50 percent noncarbonate terrigenous mud indicate that terrigenous
sedimentation has increased in the study lagoons during recent times.

9.4 Carbonate Mud

Carbonate mud covers wide areas of the open lagoon in Puerto Mosquito and
Bahia Tapon (Figs. 7 and 8). It also occurs as a wide band between Halimeda
sand and pelleted mud facies in Puerto Ferro. This facies contains broken
plates of Halimeda and scattered mulluscan debris giving a texture of sandy
mud to gravelly sandy mud. The mud fraction is primarily carbonate from the
breakdown of algae, but contains up to 40 percent non-carbonate,
terrigenous-derived mud when this facies is proximal to pelleted mud.
Scattered plant material is present when carbonate mud occurs near the
perimeter of Puerto Mosquito, Puerto Ferro, and Bahia Tapon. Bioturbation is
extensive and Thalassia is sparse. The flora and fauna typical of this
environment are Thalassia testudinum, Chione cancelata, Bulla striata, and
Cerithium variable.

9.5 Molluscan Gravel

Molluscan gravel is medium to dark gray in color with a texture ranging
from muddy, sandy gravel to gravelly, sandy mud. It contains 30 percent
mollusc shells and averages 10 Halimeda fragments in the sand fraction. The
gravel fraction typically is dominant and is comprised of mollusc shells
many of which are articulated. The mud fraction contains up to 20 percent
terrigenous, noncarbonate sediment.

When this facies occurs near the surface, the gravel and sand fractions
diminish and the unit grades into the carbonate mud facies. Field
observations indicate that live molluscs, which comprise the gravel

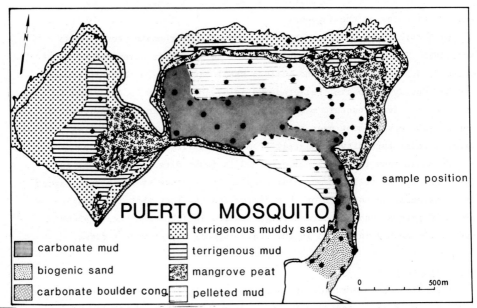

Figure 7 - Surface sediment facies distribution, Puerto Mosquito.

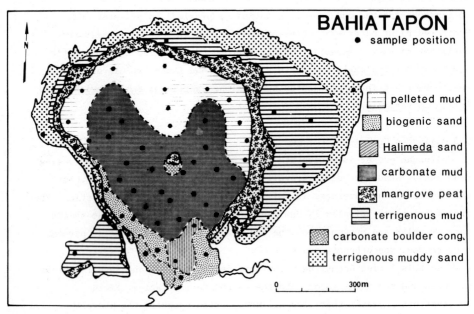

Figure 8 - Surface sediment facies distribution, Bahia Tapon.

fraction, inhabit the sediment 10 to 20 cm below the surface. They are generally absent in the upper section of this unit. The decrease in sand content can be explained by biogenically graded bedding. This grading is created through churning of sediments by Callianassa and worms. Rhoads and Stanley (1965) studied this phenomenon in Cape Cod Bay, Massachusetts, and Warme (1967) documented its development in Mugu Lagoon, California. These studies found that ample time was available to impart biological energy to sediments at the site of deposition in order to develop this grading.

Molluscan gravel is considered indicative of the moderate turbidity and sparsely vegetated environment seen in areas currently receiving carbonate mud deposition.

9.6 Halimeda Sand

Halimeda sand is restricted in the modern lagoons. Of the two lagoons detailed here, only Bahia Tapon contains the facies at the surface (compare Figs. 7 and 8). This facies is comprised primarily of nonabraded whole and broken plates of Halimeda opuntia and has a texture ranging from fine gravel to gravelly sand. Thalassia roots are common and mud is nearly absent. This lack of mud is reflected by the low turbidity of marine water over this environment in Puerto Ferro and Bahia Tapon. Molluscs common to this facies are Codakia orbiculata, Codakia orbicularis, Cerithium variabile, Anondontia alba, Bulla striata, and Tegula syba. This facies is found only in areas of low turbidity and normal marine salinity. It is also restricted to zones of moderate to low energy in the study lagoons.

Halimeda sand occurs in the cores from Bahia Tapon and Puerto Mosquito. This facies is identical in floral and faunal assemblage but differs in texture from that which is presently accumulating. The subsurface unit contains more mud and fine sand than the modern environment. A complete lack of stratification within the facies and the biological activity observed in the field imply a high degree of bioturbation as a source of this finer texture. Insoluble residue, excluding organics, represents 5 percent of the mud fraction, signifying that this unit, like its surface equivalent, received little noncarbonate, terrigenous sediment.

9.7 Mixed Skeletal Sand

Mixed skeletal sand is present in the inlets of both Bahia Tapon and Puerto Mosquito and in the southwestern margin of Bahia Tapon. These are areas of high energy, and as such they produce a well-sorted sand from the available sediment. This facies consists of fine to coarse quartz sand, Halimeda fragments, mixed shell material, coral, Homotrema, limestone fragments, and some quartz. <u>Penicillus</u>, <u>Udotea</u>, and <u>Halimeda</u> are presently living along the lower energy perimeter of this environment, and <u>Thalassia</u> is common in relatively deep water.

The subsurface facies of mixed skeletal sand was encountered in cores taken in the inlet of Puerto Mosquito and in the beach ridges of the lagoons to the east. This facies is identical in composition to its surface equivalent. Though at times better sorted than the surface counterpart, it is considered indicative of high energy, paralic environments.

9.8 Limestone Boulder

The limestone boulder facies is very poorly sorted. The boulder fraction is comprised of angular Oligocene-Miocene limestone. The remaining part of this sediment is coral and shell material from adjacent marine and lagoon environments.

Accumulations of limestone boulder border the steep slopes surrounding the southern boundaries of Puerto Mosquito (Fig. 7). The sediment here is derived primarily from the reworking of colluvium found at the base of limestone outcrops. Limestone boulder forms the southwest boundary of Bahia Tapon and the island in the center of the lagoon. This central deposit of limestone boulder is considered indicative of very high energy storm events. Evidence includes similar nearby accumulations of coral and limestone boulders, which natives attribute to recent hurricane activity. The exact thickness and extent of this unit is not known because coring was not possible due to the coarse nature of the sediment. The surface and shallow subsurface occurrences of this facies are equivalent in composition.

9.9 Gravel Deposits

The gravel deposit facies, similar to the limestone boulder facies, is created by storm activity. Its composition depends on the sediments present in adjacent environments at the time of the event. These units are recognized by the sharp contact with the facies directly beneath and by a high gravel content, commonly over 50 percent. Mollusc shells are abraded and disarticulated as compared to articulated, nonabraded shells found in other lagoonal facies.

Gravel deposits range from muddy, gravelly sand to sandy gravel, with various molluscs, <u>Halimeda</u>, and fragments of <u>Siderastrea</u> comprising most of the gravel fraction. The coarse particles are both reworked from older lagoonal deposits and transported into the lagoon from the reef environments that border Vieques on the south.

No recent gravel deposits were observed on the surface of the present lagoon, but an outcrop of a large lens was found at the inlet of Bahia Tapon. This outcrop consists primarily of branches of Siderastrea. Gravel deposits were encountered in cores from both Bahia Tapon and Puerto Mosquito. These deposits contain over 50 percent gravel, primarily mollusc shells and Siderastrea. The remainder of the deposits was of mixed composition from adjacent environments.

9.10 Seagrass Peat

Seagrass peat occurs in beds 0.1 to 0.4 m thick. The facies can be totally peat or may appear as thin beds interlayered with molluscan gravel of carbonate mud. This peat is dark red-brown in fresh sample and black when oxidized. It is comprised mainly of roots and blades of Halodule wrighti, with sporadic blades of Thalassia testudinum. Seagrass peat is more dense by mass than mangrove peat and shows crab burrows through 20 percent of the cross-sectional areas. These burrows are filled with material from the overlying sediments. As in the mangrove peat, some zones are muddy. Where mud occurs, it is a mixture of terrigenous silt and clay and some carbonate mud. Sand-sized fragments of Halimeda plates, mollusc shells, and quartz are common in thin beds.

9.11 Neritina mud is brownish-black to black in color and contains abundant organic material. It is typically a gravelly, sandy mud to a gravelly, muddy sand. The gravel fraction is primarily gastropod shells. The sand fraction contains quartz, epidote, and fragments of Halimeda plates. Halodule roots are found scattered irregularly within the clayey matrix. Neritina virginea and Neritina punctulata are prolific and comprise 90 percent of the gravel fraction. Cerithium variabile is also present. Chione cancellata is found sporadically in the upper part of this unit.

No living Neritina individuals were observed during the field work; however, Abbott (1974) describes their habitat as upper intertidal brackish flats and occasionally hypersaline areas. Consequently, this facies is considered to be indicative of brackish to hypersaline intertidal mud flats.

10 DESCRIPTION AND INTERPRETATION OF STRATIGRAPHIC CROSS SECTIONS

10.1 Because the Holocene sea-level rise controlled the depositional history of Vieques' lagoons, it will be discussed briefly in order to set the scene for stratigraphic interpretations.

No work has been done on the observable effects of sea level on the island of Vieques itself, but the proximity and close structural association of this island with the main island of Puerto Rico permits these two islands to be considered as one unit. Kaye (1959) has conducted an in-depth study of Puerto Rican shoreline morphology in order to delineate changes in sea level for the Holocene. In his study, the more recent part of this record was constructed using beach rock, wave-cut notches, tidal terraces, and beach ridges.

Kaye (1959) determined that sea level began to rise 4000 yr B.P., drowning eolianites that were formed during a lower stand abut 4 to 8 m below the present level. Scholl (1964a, b) and Scholl and Stuiver (1967) studies sea-level changes on the south coast of Florida and determined that sea level rise slowed during the last 3-5000 years of the Holocene, but that it has continued to rise to the present. A similar pattern has been found by Shepard (1960), who studied this phenomenon on the northwest coast of the Gulf of Mexico, and by Bloom (1970) in a study of the stratigraphy of the eastern Caroline Islands.

A rise in sea level is reflected in the stratigraphy of the lagoons of Vieques and will be obvious in the succession of facies described in the following section. A ^{14}C radiometric age of 560 ± 50 yr.B.P. has been determined for a sample of mangrove peat taken 0.4 m below mean low tide, suggesting that the rise in sea level exhibited in the stratigraphy of the lagoons of Vieques is of a recent nature, approximating that proposed by Scholl (1964a), Shepard (1960), and Bloom (1970).

10.2 Stratigraphy

Stratigraphic cross sections of the four study lagoons have been developed using information derived from cores. Cross sections of Bahia Tapon and Puerto Mosquito are presented in Figure 9.

The cross section of Puerto Mosquito (Fig. 9A) shows the ideal succession of facies. As in all of the lagoons, Puerto Mosquito is dominated by terrigenous alluvium at the landward end of the cross section (Fig. 9A). Alluvium is oxidized at the surface and is partially reduced near the bottom of core PM-1. This core also contains terrestrial plant root material. Chemically reduced terrigenous alluvium is present at the base of cores PM-1, 2, 4, and 5, and is interpreted as representing subaerial deposition of alluvium prior to and during Holocene marine flooding.

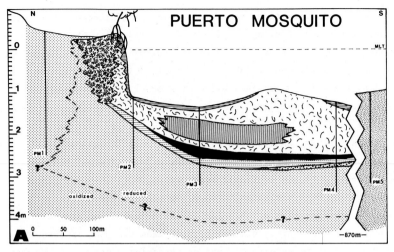

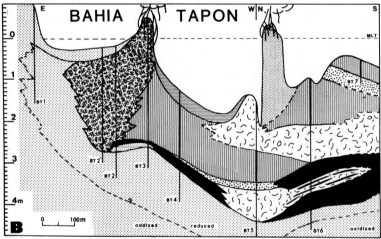

Facies Key to Stratigraphic Cross Sections

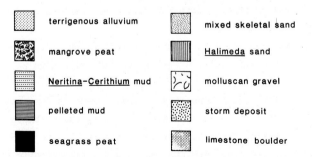

terrigenous alluvium		mixed skeletal sand
mangrove peat		Halimeda sand
Neritina-Cerithium mud		molluscan gravel
pelleted mud		storm deposit
seagrass peat		limestone boulder

Figure 9 - Stratigraphic cross-sections at (A) Puerto Mosquito and (B) Bahia
Tapon showing distribution of sedimentary facies. Note that
vertical exaggeration is high and that it is about twice as much
for Bahia Tapon as for Puerto Mosquito.

As sea level rose, storm tides deposited a mixture of abraded shell and coral on the terrigenous alluvium (Fig. 9A; core PM-4). When the sea flooded Puerto Mosquito to an intertidal to slightly subtidal level, Nertina mud was deposited over the terrigenous alluvium and storm facies (Fig. 9A; cores PM-2, 3, and 4). This facies is indicative of a mud-flat environment. With rising sea level, mud deposition transgressed toward the periphery of Puerto Mosquito, and a less extensive unit of seagrass (Halodule) peat was deposited in the subtidal central areas. Seagrass peat is indicative of shallow, subtidal, marine water with polyhaline and polythermal conditions.

As Puerto Mosquito became subtidal, mangrove colonization took place along the shores. No core was taken through the peat on the northern end of Puerto Mosquito. The position of this facies was interpolated from surface exposure and from the stratigraphic position of mangrove peat in other lagoons. Rising sea level appears to have exceeded sedimentation rates at this point in the history of Puerto Mosquito because facies indicative of deeper, more open lagoonal environments were deposited next. This condition is represented by molluscan gravel, a facies indicative of turbid marine water and limited terrigenous sedimentation. Molluscan gravel was deposited over the entire open lagoon until it was interrupted along the northern third by Halimeda sand (Fig. 9A). The presence of Halimeda is indicative of good circulation and low turbidity. Together, these two facies represent the major marine environments of lagoon deposition. The cross section appears to signify contemporaneous existence of Halimeda sand and molluscan gravel for a relatively long depositional period. According to modern restrictions of Halimeda sand, circulation must have been good and turbidity low for the Halimeda sand facies to be deposited. Possibly the tombolo to the west of Puerto Mosquito had not yet formed, and this situation afforded better circulation. Halimeda sand deposition was replaced by molluscan gravel during the later development of Puerto Mosquito, representing a return to turbid conditions within the lagoon.

The next change in deposition is relatively recent and is represented by accumulation of pelleted mud. This facies is composed of a high percentage of terrigenous sediment and is indicative of a high rate of erosion in the surrounding terrain. The modern depositional environment is dominated by molluscan sand in central areas and pelleted mud along the northern and southern boundaries. During deposition of marine facies in the open lagoon, mangroves continued to prograde into the lagoon and until recently they restricted terrigenous sedimentation to the alluvial flats.

The landward boundary of Bahia Tapon is composed of terrigenous alluvium (Fig. 9B). This facies is chemically reduced in cores BT-2'. 2. 3. 4. 5. and 6, and the lower part of core BT-1. As in the other lagoons, this unit contains terrestrial plant root material in situ and is indicative of terrestrial deposition.

No storm deposit was found directly above the terrigenous facies as in Puerto Mosquito, but this may be due to shielding provided by outcrops of limestone on the west side of the inlet.

Bahia Tapon appears to have had a mixed set of environments in its early history because the facies of Neritina mud, seagrass peat, and molluscan gravel appear to have been deposited contemporaneously.

Neritina mud, which represents a mud flat environment, was deposited above the relatively horizontal zone of terrigenous alluvium and was penetrated by cores BT-2 and 3 (Fig. 9B). As previously discussed, sea level or the original slope of terrigenous alluvium appear to have controlled the position and extent of this facies.

The slope of terrigenous alluvium seems to approximate paleotopography, the center of Bahia Tapon being deeper and the periphery more shallow and suitable for accumulation of seagrass peat. Molluscan gravel was deposited in the deeper central area. Core BT-6 (Fig. 9B) penetrated beds of seagrass peat interbedded with molluscan gravel, indicating that the environment within Bahia Tapon changed often, possibly due to storm deposition near the inlet, restricting normal marine circulation. An extensive storm facies encountered in the open lagoon in cores BT-4 and 5 (Fig. 9B) appears at a contemporary stratigraphic level. Mangrove colonization took place along the shoreline and deposition of mangrove peat began during the early subtidal history of Bahia Tapon.

Depositional environments changed and deeper water facies became dominant with rising sea level. Molluscan gravel was deposited on the southern half of Bahia Tapon, and while Halimeda sand accumulated in the northern half. Within the cross section (Fig. 9B) it appears that deposition of Halimeda sand was halted by a transgressing molluscan gravel facies, which in turn was superseded by Halimeda sand in the southern end. Storm deposition and eolianite exposures at the surface are probable factors contributing to the changing conditions displayed in more recent lagoon history. Outcrops near the inlet presently protect Bahia Tapon from heavy wave activity, and storm deposition near the mouth of the lagoon restricts circulation of marine water.

The deposition of _Halimeda_ sand in the southern part of Bahia Tapon was later interrupted by a storm deposit composed of branches of _Siderastrea_ (Fig. 9B). Sedimentation of Halimeda sand then resumed and has continued to the present. _Halimeda_ sand in the northern half of Bahia Tapon gives way to recent deposition of pelleted mud (Fig. 9B) and indicates an increase in the erosion of surrounding terrain similar to that seen in the Puerto Mosquito.

11 SUMMARY OF LAGOON STRATIGRAPHY

The early stratigraphic history of the lagoonal basins is similar. The first sediments received were deposits of terrigenous alluvium from the erosion of the main igneous body of Vieques. As rising sea level "flooded" the lagoons, mud flat facies were deposited. Seagrass peat accumulated in adjacent subtidal zones, where conditions of temperature and salinity varied.

Deposition of marine derived sediments indicative of better circulation and deeper water began after deposition of _Neritina_-_Cerithium_ mud and seagrass peat. Puerto Mosquito, having the poorest circulation, has been dominated by deposition of mulluscan gravel to the present, indicating a more turbid, shallow environment receiving limited amounts of terrigenous sediments. Bahia Tapon, having better circulation, displays larger deposits of _Halimeda_ sand. From the relative stratigraphic positions of these facies, Bahia Tapon also experienced a more mixed and complex set of conditions throughout its history. The modern depositional environments of the lagoons show a unit of pelleted mud that contains about 50 percent non-carbonate terrigenous sediment along lagoon margins. A high rate of erosion of the surrounding terrain and/or a loss in population of surrounding mangroves is indicated. Increased terrigenous sedimentation has been attributed to the construction of unpaved roads and overgrazing of cattle (Lewis et al, 1981).

All four lagoons studied on Vieques exhibit a similar stage in sequential development and, as a result, only an outline of possible future stages can be theorized. Using the present trend of increasing terrigenous sedimentation and mangrove progradation, these lagoons may become dominated by such sediment. The lagoons may remain open to marine circulation or, due to storm deposition or increased erosion, they may become restricted and form larger hypersaline flats similar to the areas of the alluvium behind fringing mangroves.

Because colonization by the seed of _Rhizophora mangle_ is restricted to 0.5 m water depth (Wanless, 1975), the depth of lagoon perimeters is a controlling factor in mangrove population shifts. Water depth in Puerto Ferro is commonly less than 0.5 m along the mangrove fringe (Fig. 5), and

Rhizophora has shown considerable colonization into the lagoon over the last 40 years. The average margin of Puerto Mosquito is deeper than that of Puerto Ferro, and air photos show no major shift in mangrove population since at least since 1936. The mangrove population of Bahia Tapon has encountered a similar barrier with peripheral depths greater than 0.5 m restricting colonization.

Continued sedimentation on alluvial flats has caused losses in the landward populations of fringing mangroves by raising an increasing area of the flat to a subaerial level. Losses in landward populations are also caused by the stress of changing salinities in hypersaline pools. If these current trends continue, losses in the fringing populations of mangrove will be reduced to a thin fringe in Puerto Mosquito and Bahia Tapon until lagoonal areas become shallower and more suitable for colonization.

Although each of the lagoons in the study is small, it is typical for high-relief islands such as Vieques or others in the Caribbean, to have many such environments along their irregular coast. The rather high relief provide a supply of terrigenous sediment. The open marine environment adjacent to this irregular coast is typically a reef dominated carbonate system. Detrital carbonate sediments from the reef complex combine with autochthonous biogenic carbonate sediment in the lagoons. The result is an interfingering of terrigenous derived siliciclastic facies with marine derived carbonate facies.

The preservation potential of these well-protected lagoonal sediments is rather high especially in a tectonically active area such as the Caribbean. Similar situations exist in the Pacific Ocean as well. As a result the stratigraphic record produced in this setting would be expected to have a relatively extensive marine carbonate component with local accumulations of carbonate-terrigenous transition sediments. The latter represent coastal lagoons similar to those on Vieques. Presence of such transition facies would be of great importance in reconstructing the paleogeography and depositional environments in which these transition facies accumulated.

ACKNOWLEDGMENTS

This paper is taken from the M.S. thesis of the senior author. The financial support of Mangrove Systems, Inc. through a contract with the U.S. Department of the Navy is gratefully acknowledged. Additional support was provided by Sigma Xi, the Scientific Research Society. Field work benefited from the assistance of Jim Derrenbecker, Robin Lewis, Ralph Lombardo and Tom Stubbs. R.P. Wunderlin identified plant debris in the peats. Ralph Levingston designed and constructed the coring apparatus.

REFERENCES

Abbott, R.T., 1974. American Seashells: New York, Van Nostrand-Rheinholt, pp. 656.

Benitez, J.A.B., 1976. Vieques en la historie de Puerto Rico: San Juan, F. Ortiz Nieves, table 7, 122.

Bloom, A.L., 1970. Paludal stratigraphy of Turk Ponape and Kusiae, eastern Caroline Islands: Geol. Soc. America Bull., 81, 1895-1904.

Briggs, R.P., 1964. Provisional geological map of Puerto Rico and adjacent islands: U.S. Geol. Surv., Misc. Geol. Inf., Map 1, p. 392.

Davis, J.H., Jr., 1940. The ecology and geologic role of mangrove in Florida: Carnegie Inst. Washington Publ. 571, pp. 303-412.

Fairbridge, R.W. (Ed.)., 1975. Encyclopedia of World Sciences, Stroudsburg, Pa., Dowden, Hutchinson and Ross, Inc., pp. 434-437.

Kaye, C.A., 1959. Shoreline features and Quaternary shoreline changes, Puerto Rico: U.S. Geol. Surv. Prof. Paper, 317-B, pp. 49-140.

Lewis, R.R., Lombardo, R., Sorrie, B., D'Aluisio-Guerrieri, G.M. and Callahan, R., 1981. Mangrove forests of Vieques, Puerto Rico: Report to the U.S. Navy, Tampa, Florida, Mangrove Systems, Inc. 1, p. 41.

Rhoads, D.C. and Stanley, D.J., 1965. Biogenic graded bedding: Jour. Sed. Pet., 35, 4, 956-963.

Scholl, D.W., 1964a. Recent sedimentary record in mangrove swamps and rise in sea level over the southwestern coast of Florida, Part 1: Mar. Geol., 1, 344-364.

Scholl, D.W., 1964b. Recent sedimentary record in mangrove swamps and rise in sea level over the southwestern coast of Florida, Part 2: Mar. Geol., 2, 343-364.

Scholl, D.W. and Stuiver, M., 1967. Recent submergence of southern Florida: a comparison with adjacent coasts and other eustatic data: Geol. Soc. America Bull., 78, 437-454.

Shepard, F.P., 1960. Rise of sea level along the northwest Gulf of Mexico: In: Shepard et al. (Eds.), Recent Sediments, Northwest Gulf of Mexico: Am. Assoc. Petrol. Geologists, Tulsa, Oklahoma, pp. 338-344.

Wanless, H.R., 1975. Mangrove sedimentation in geology perspective: In: Gleason, P.J. (Ed.), Environments of South Florida, Present and Past: Miami Geol. Soc., Mem. 2, pp. 190-200.

Warme, J.E., 1967. Graded bedding in recent sediments of Mugu Lagoon, California: Jour. Sed. Pet., 37, 2, 540-547.

Warme, J.E., 1969. Live and dead molluscs in a coastal lagoon: Jour. Paleontology, 43, 1, 141-150.

Woodbury, R.O., 1972. Vieques 1972 survey of the natural resources, sec. VII, part VII: Rio Piedras, Universidad de Puerto Rico, pp. 1-16.

Chapter 8

CARBONATE-TERRIGENOUS SEDIMENTATION ON THE NORTH PUERTO RICO SHELF

O.H. PILKEY, D.M. BUSH Duke University, Department of Geology, Durham, NC
27708
R.W. RODRIGUEZ
U.S. Geological Survey, San Juan, Puerto Rico   00906

ABSTRACT
    The narrow steep high wave energy north shelf of Puerto Rico is
characterized by rapid terrigenous sedimentation off river mouths and slow
accumulation of purely calcareous non-reefal sediment between river mouths.
Fine grained terrigenous sediments accumulate in pods or wedges up to 40
meters in thickness. Shifting of river mouths moves the loci of terrigenous
sedimentation which is then replaced by carbonate sedimentation. After
abandonment of a river mouth location, the accumulated offshore terrigenous
sediment, if located near one of the many canyons indenting the shelf, is
gradually removed from the shelf, down the canyon system. It is believed
likely that transgressions of sea level remove any accumulated terrigenous
sands seaward down the steep shelf. Hence the present day accumulations and
lateral relationships of carbonate non-carbonate depocenters on the north
Puerto Rico shelf have largely formed during the Holocene.

1 INTRODUCTION

    The north shelf of Puerto Rico is steep and narrow, and is subjected to

trade wind dominated high wave energies. Most of the land area of the island

drains to the north shelf where river mouths are spaced at ·5 to 10 mile

intervals. Fluvial sediments arrive on the shelf almost exclusively as a

result of hurricane related flood events. Because the shelf is so narrow and

steep, sediment dispersal tends to be perpendicular to the shoreline and

lateral transportation seldom extends beyond a few miles from a river mouth.

As a consequence, the Puerto Rico north shelf is characterized by an unusual

surficial sediment cover: sharply defined patches of terrigenous sediments

coexisting with areas of solely calcareous sediment (Schneiderman et al.,

1976, and Pilkey et al., 1978). The terrigenous sediment has the added

unusual characteristic of existing in equilibrium with present-day physical

processes on the insular shelf.

    The purpose of this paper is to summarize our understanding to date of

the insular shelf and to present a sedimentation model for it that

emphasizes the interrelationship between the carbonate and noncarbonate

sediment. Basically, the northern shelf of Puerto Rico has a continuous but

slow accumulation of nonreefal calcareous materials together with a shifting

overprint of rapidly deposited terrigenous materials. The model developed

should have application to most windward insular shelves on Caribbean and

232

other islands, although on many of those shelves reefs would be a factor contributing to higher rates of carbonate sedimentation.

More important, however, is that the northern shelf of Puerto Rico can be construed as a geologically compressed system, an understanding of which should have application beyond insular shelves. Because of the narrow shelf and the lack of sediment-trapping estuaries, cross-shelf processes occur with great rapidity compared to those on broader continental shelves. Flood sediment is dispersed across the entire shelf and arrives at a textural equilibrium stance within months (Grove et al., 1982). Because so many of the world's modern clastic shelf environments have not yet attained equilibrium with the physical environment and are covered by relict sediments, the understanding of shelf processes in an environment like the Puerto Rico north shelf may offer a next best alternative to actual study of nondeltaic continental shelves.

2 STUDY AREA SEDIMENTARY FRAMEWORK

Puerto Rico is a rectangular island about 160 km long and 50-60 km wide (Fig. 1). An east-west mountain belt, the Cordilleran Central, provides the igneous and highly folded meta-sedimentary backbone of the island. Tertiary limestone formations and Quaternary terrigenous and calcareous unconsolidated sediments flank the mountains and, in most cases, underlie the narrow coastal plains (Briggs and Akers, 1965). The crest of the Cordilleran Central is located close to the southern coast and because of the northerly tradewinds, the largest rainfall is received on the northern part of the island (Lopez and Colon-Dieppa, 1973). Thus, the rivers crossing the northern coast are the island's largest; because they drain most of the island's land area, the north shelf receives by far the largest terrigenous load (Ehlman, 1968). The north shelf ranges in width from 0.5 to 10 km, averages about 3 km, and breaks at about 70m water depth. A number of submarine canyons indent the outer shelf.

The north shelf is wave dominated and is subjected to open-ocean waves formed by the tradewind system and to swells from North Atlantic storms. Because the Puerto Rico shelf is steep, narrow, and straight, storm surge and associated shelf currents are minor. Storm-wave swash is a more important coastal process and according to Fields and Jordan (1972), North Atlantic hurricanes off the east coast U.S. often cause major storm swash events on the north Puerto Rico coast.

Few physical oceanographic studies have been made on the Puerto Rico north shelf and what data are available are in the "gray literature" (e.g.

233

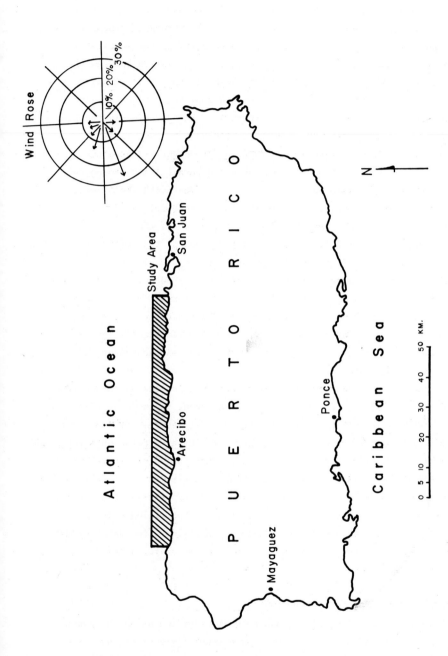

Fig. 1. Index map showing location of the study area in Puerto Rico, and a wind rose from San Juan (from Pilkey et al., 1978).

Puerto Rico Department of Public Works, 1974). Tidal amplitude is less than 0.5 meter. Dominant surface-current movement on the north shelf is to the west. According to Wood et al. (1975), "The usually strong afternoon winds from the east-northeast tend to increase the velocity of the surface currents to the west." Wood also notes "a strong correlation between the current patterns and the tides with modification by the local winds, the North Equatorial Current and the direction and amplitude of sea swells impinging on the shoreline." Bottom currents have been studied in the Rio de la Plata Submarine Canyon by Shepard et al. (1979).

The narrow Puerto Rico insular shelf has a very diverse and patchy sediment cover, reflecting the wide variation in physical and biological factors affecting the shelf environment (Fig. 2). The north shelf, which is subjected to the highest wave energies and the largest influx of river sediment of all Puerto Rico shelf segments, has no coral reefs (Schneidermann et al., 1976). Sediment cover on much of the north insular shelf is in textural equilibrium with present day physical processes. Cross-shelf patches of dark-colored (and hence easily identifiable) river-derived muddy sand are found adjacent to every river mouth. These patches are separated by regions of light-colored medium to coarse sand dominated by mollusks, foraminifera, and calcareous red algae fragments. Boundaries between the two major sediment types, one allochthonous and the other autochthonous, are very sharp.

The portion of the north shelf studied in detail is located between Rio Camuy to the west and Rio de la Plata, near San Juan, to the east (Fig. 1). The main database of this study is a series of grab samples obtained over the last several years using the research vessels Cape Hatteras and Jean A. Our conclusions are also based on preliminary study of 30 vibracores and piston cores obtained mostly off the Rio Grande de Arecibo and the Rio de la Plata river mouths. High resolution, shallow seismic profiles on a 200 meter spacing were obtained as part of a nuclear power plant siting survey (Western Geophysical, 1975) and provide an unusually detailed database for interpretation of the upper sediment column (See Fig. 3 for sample locations).

3 CARBONATE AUTOCHONOUS SEDIMENTATION

Most of the surficial shelf regions between river mouths on the Puerto Rico shelf are covered by a thin veneer of essentially pure calcareous sands. This sand is generally white to light brown in coloration, in strong contrast to the dark color of terrigenous sands derived from rivers.

235

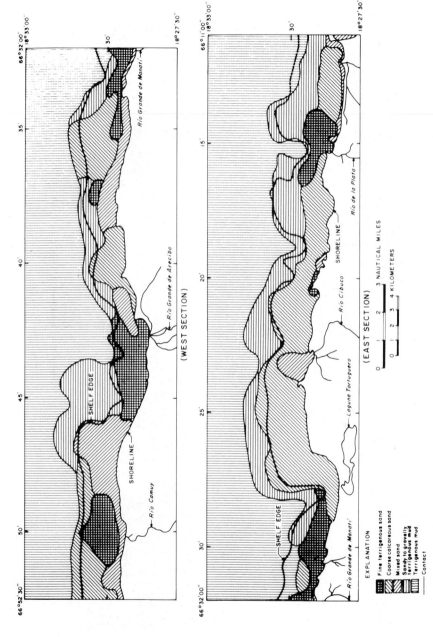

Fig. 2. Bathymetry (shelf edge) and surficial sediment types of the North Shelf study area.

236

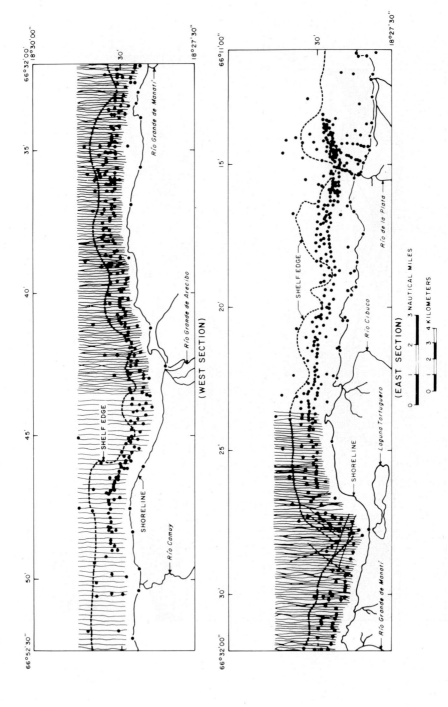

Fig. 3. Map showing tracklines of seismic reflection profiles obtained by Western Geophysical Company (1975) and sample locations.

Mollusks are the dominant skeletal constituent. Halimeda plates, foraminifera, coral fragments, bryozoans, barnacles, and red algal fragments are of varying but secondary importance. In places, rhodoliths (red algal nodules) as much as 3 cm or greater in diameter form more than 75% of the sediment cover (Fig. 4A). In all cases the rhodoliths are now found in water depths greater than 30 meters although they very likely formed in much shallower water (Adey and MacIntyre, 1973).

Individual particles in the carbonate sediment are characterized by a high degree of fragmentation; there is much evidence of physical rounding and of biodegradation, particularly through the activities of microboring organisms (Fig. 4B). On the basis of displaced faunal constituents, the presence of rhodoliths, and two total sediment radiocarbon ages of 680 ± 50 and 2530 ± 90 years BP (Stout, 1979) for middle and outer-shelf calcareous samples, respectively, it is apparent that the carbonate sediment is relict or autochthonous. Accumulation of calcareous sediment in this nonreef environment is obviously very slow. The strongest indication of this is the mixture of modern and relict material in the surficial layer.

The physical condition of beach and nearshore calcareous sands differs significantly from that of shelf sands. Beach sands tend to be highly rounded (Fig. 4C) to the extent that individual components are seldom identifiable except by thin section examination (Pilkey et al., 1979). More important, beach derived grains are highly polished. In the dynamic environment of the high energy beaches along Puerto Rico's north shore, the physical abrasion processes that polish override the surface-dulling activities of microboring organisms. Occasional, highly polished, calcareous grains in inner and middle shelf sands clearly point to the contribution of the beach environment. Such grains are most commonly preserved in the carbonate fraction of rapidly deposited storm layers in the terrigenous sedimentation areas.

4 TERRIGENOUS ALLOCHTHONOUS SEDIMENTATION

The sediment cover off river mouths is entirely modern, still connected to its fluvial source and in equilibrium with present day physical processes. The sediment cover here is allochthonous according to the terminology of Swift (1976).

Typically the color of this sediment type is black to olive-gray, in strong contrast to the light gray to yellow-brown color of calcareous sediments. Primary constituents of terrigenous sediments are volcanic rock fragments and quartz-feldspar. Magnetite and hornblende are the principal heavy minerals (Pilkey and Lincoln, 1984). Other heavy minerals include a

238

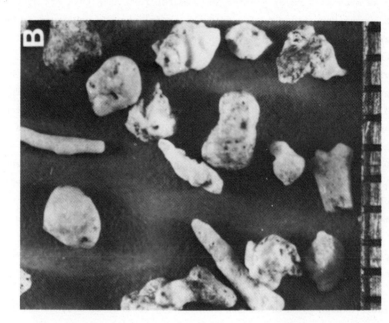

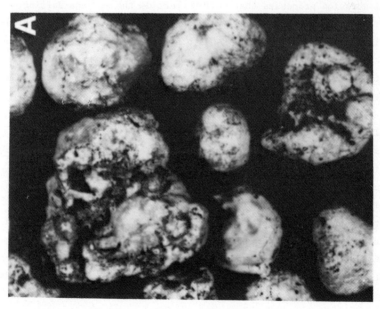

Fig. 4. Photographs of coarse carbonate fractions from various sediment types. A) Rhodolith dominated sample (true scale). B) Typical shelf carbonate showing degradation of grains (scale in mm).

Fig. 4. Photographs of coarse carbonate fractions from various sediment types. C) Beach sand showing polish of grains (scale in mm). D) Coarse carbonate fraction from a terrigenous-dominated sand (scale in mm).

variety of pyroxenes, epidote, monazite and some secondary minerals such as limonite, hematite, and pyrite. The calcium carbonate content of this sediment type is highly variable, depending on nearness to river mouths, but it averages about 28%. Mean grain size is commonly in the fine sand range.

Calcareous materials are, of course, also being added to the terrigenous sediment patches. The carbonate content of terrigenous sands ranges from 10% or less at river mouths to 50% or more in boundary zones where the two sediment types are mixed. The carbonate fraction of terrigenous sands is fresh in appearance, often with original coloration (Fig. 4D).

The surficial sediment cover of the allochthonous, terrigenous sands exhibits clear cross-shelf trend of si.e grading. At depths shallower than 40 meters the surficial sediment is fine sand with only minor amounts of mud. The sediment becomes increasingly muddy below 40 meters to the shelf break at 75 meters. There apparently is little onshore transfer of sand from the shelf to beaches or nearshore zones. Immediately in front of river mouths, the terrigenous sand band is continuous into shallow water. Away from river mouths, however, the inner shelf to depths of 20 meters is largely a rocky abrasion platform with occasional patches or a thin veneer of unconsolidated calcareous sediment. The exposed rock consists principally of well lithified calcareous aeolianite of Pleistocene age. Sediment patches within the narrow, high energy, inner-shelf zone rarely contain significant amounts of fluvially derived grains, even when the adjacent middle and outer shelf are covered by sediment of entirely terrigenous grains.

The inner shelf boundary between the terrigenous and carbonate sands is usually very sharp; occurring within a distance of a few tens of meters. On the mid-shelf, however, the terrigenous-sediment bands grade into carbonate sands through a zone of mixed sediments (Fig. 2). The zone of mixing may be between several hundred meters to one kilometer in width. The mixed sediments have a characteristic salt-and-pepper appearance due to the addition of light colored calcareous material, which also is responsible for a mean grain size larger than that of the purely terrigenous sand.

5 SURFICIAL SHELF SEDIMENTATION PROCESSES

Two types of storms affect the north shelf of Puerto Rico. Hurricanes that actually strike Puerto Rico result in both intense wave activity and river flooding. Such storms result in deposition of inundites according to the terminology of Seilacher (1982). Extratropical storms result in wave activity without accompanying floods producing tempestites (Seilacher, 1982). Such storms play a role in redistribution and perhaps destruction of previously deposited inundites; greatly increased amounts of suspended material has been observed in the shelf water column during storms.

Using distinct heavy mineral suites to trace terrigenous sediment movement and mixing has helped to formulate our sedimentation model (Grossman, 1978). River sand entering the shelf system is generally transported seaward to a depth of 15 to 20 meters. After which the sediment either moves laterally both to the east and west (Fig. 2) or continues seaward into the submarine canyon system present at each river mouth. The relative importance of down-canyon movement off the Puerto Rico shelf compared to lateral movement on the middle shelf is not known.

Fuerst (1979) observed graded beds of terrigenous sands on the upper insular slope in the vicinity of the Rio de la Plata canyon system. However, the Holocene/Pleistocene sediment cover is thin on the slope, indicating that deposition here is shortlived before removal to deeper water by mass transport processes down the steep landward slope of the Puerto Rico trench.

Cross-shelf transport of mud was studied off the Rio de la Plata river mouth (Grove et al., 1982). After the passage of Hurricanes David and Frederick in 1979, shelf sampling revealed the presence of widespread mud and muddy sand in depths shallower than 40 meters. Within six months the muddy sediment had been entirely removed, presumably by subsequent minor storm activity. The process of cross-shelf transport of mud after major storms has been termed "mud hopping" (Pilkey et al., 1978). Deposition of the mud occurs initially for two reasons: first, the instantaneous injection of a large mass of sediment probably exceeds the transportation capacity of shelf processes to quickly move the sediment across the shelf. Perhaps more important is the fact that the river floods and, hence, the arrival of the sediment to the shelf occurs after the hurricane has passed, when fairweather sea conditions once again prevail.

The thickness of the unconsoliated sediment cover on the north shelf is shown in Figure 5. These thicknesses, which are estimated from two-way seismic travel time, may be as great as 40 meters. Particularly well developed wedges are present off the two largest rivers, Manati and Arecibo.

The thickness distribution clearly shows the lateral movement of fluvial sediment on the mid-shelf away from the river mouths (Fig. 5), which was also reflected in the areal distribution of surficial sediment (Fig. 2). Although the portion of the sediment remaining on the shelf moves both east and west from the river mouths, it is clear from Fig. 5 that the net westward movement is greater. Lateral transport of large columns of sand is restricted to within five nautical miles of the fluvial sources, although minor amounts of river sand may travel much further, especially on the shelf area between the Manati and Arecibo river mouths.

242

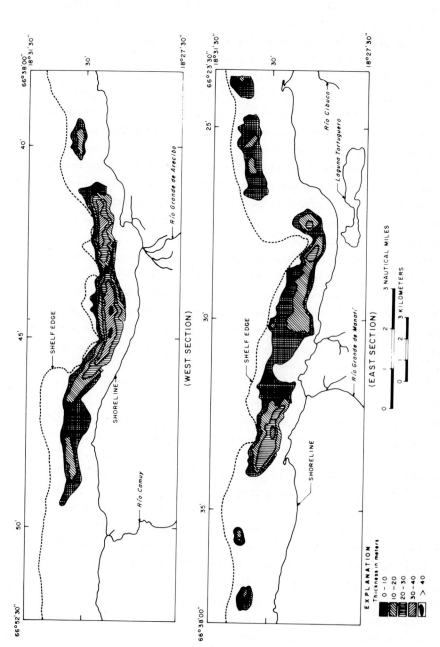

Fig. 5. A plot of approximate thickness of unconsolidated sediment cover on a portion of the North Shelf study area. Thicknesses are based on the fact that two-way travel time is approximately equal to one meter of thickness (from Western Geophysical, 1975).

Preliminary examination of vibracores taken in February, 1984 off Rio Grande de Arecibo and the Rio de la Plata indicate the presence of individual storm layers, 10-20 cm thick. In some cases the layers appear to be graded and separated from one another by coarse lag layers of shell material. The lag layers could either be of fair weather origin or may have formed as a result of winnowing at the onset of the storm which ultimately caused introduction of new material to the shelf.

Figures 6 and 7 are interpretations of high resolution seismic lines from different shelf areas. Figure 6 is a shore perpendicular line off the Rio Grande de Arecibo. Figure 7 also is a shore perpendicular line, off Laguna Tortuguero.

Figure 6, in an area where the surficial sediment type is terrigenous muddy sand, clearly shows the modern wedge of material that is prograding over the "basement" of tertiary limestone. Maximum thickness of allocthonous sediments on this line is 40m. This is based on the fact that 1 millisecond of two-way travel time is approximately equal to one meter of thickness.

Figure 7 is an interpretation of a seismic line in an area where the surficial sediment type is totally calcareous, but with a great deal of unconsolidated sediment accumulated (see also Figure 5 off Laguna Tortuguero). Clearly shown is an old river channel that was filled by a rapid sedimentation process. This is evidenced by the chaotic seismic facies in the channel. There is no river presently draining into this shelf area. Presumably the Rio Cibuco, now 10km to the east, once drained here and was responsible for the channel cut and fill.

The major questions in this area are: (1) exactly when did terrigenous sedimentation cease and carbonate sedimentation commence, and (2) how thick is each facies? The answers to these questions will hopefully come as a result of vibracoring planned for Spring and Summer, 1985. However, from observations of carbonate sedimentation rates and of cores on other portions of the shelf, it is unlikely that the surficial calcareous sediments extend to depths of greater than 10's of centimeters.

6 A MODEL FOR CARBONATE TO NONCARBONATE SEQUENCES ON A NARROW INSULAR SHELF

It is apparent that terrigenous sediments are accumulating on this narrow insular shelf at a much higher rate than are carbonate sediments. A variety of evidence points to this conclusion, including the known high frequency of sediment-laden flood waters reaching the north shelf and seismically determined sediment thicknesses ranging from over 40 meters for some terrigenous sands to less than 10 meters (commonly less than 1 meter) for carbonate sands (Fig. 5).

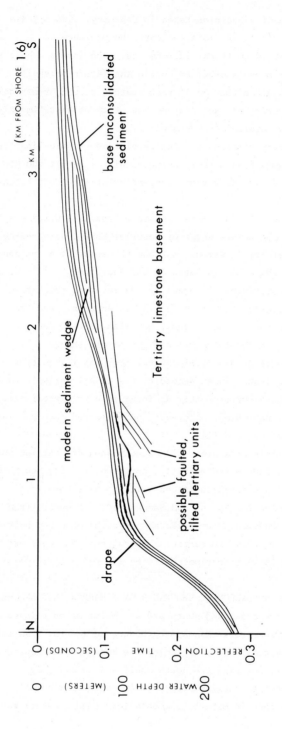

Fig. 6. An interpretation of a typical seismic profile across an area of terrigenous sediment accumulations. In this profile, directly off the mouth of the Rio Grande de Arecibo (Fig. 5), the "hardrock" basement of Tertiary Limestone is overlain by Holocene/Pleistocene (?) unconsolidated shelf sediment which has prograded out or draped over the uppermost insular slope. The modern sediment in this profile is fine grained terrigenous sediment which becomes increasingly muddy in a seaward direction. Profile is Western Geophysical (1975) Sparker line #223.

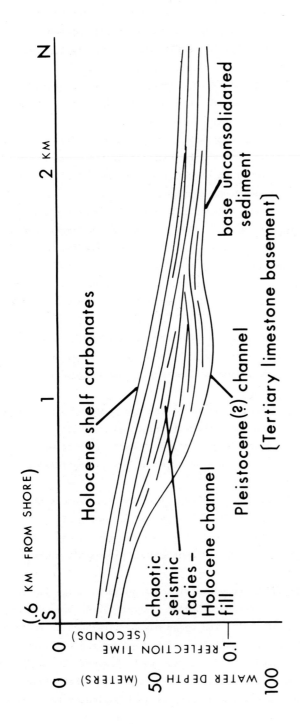

Fig. 7. An interpretation of a typical seismic profile across an area of carbonate sediment accumulation. In this profile just east of the canyon off Laguna Tortuguero, the Tertiary Tortuguero, (Fig. 5), the Tertiary Limestone "basement" shows a channel that was presumably cut during a lower Pleistocene sea level stand and that most likely connects up to the large canyon system further offshore (Fig. 2). The chaotic-fill seismic facies is indicative of rapidly deposited, probably terrigenous sediments. Presently, no river empties onto the shelf in this area, and thus, the surficial sediment type is largely calcareous. Probably only a thin veneer of carbonate material is present, however, with older Holocene and Pleistocene (?) terrigenous sediments underlying. This profile is Western Geophysical (1975) Sparker line #45.

If a standstill of sea level and/or a slow subsidence of the insular
shelf are assumed, the cross-shelf terrigenous wedges should spread
laterally and eventually cover the entire shelf. This is due to the shifting
of river mouths, a process which has characterized the Holocene-Pleistocene
history of the narrow Northern Puerto Rico coastal plain. Topographic
expressions of river channels abound on the lower coastal plain. In some
instances, it is not possible to discern which river mouth was responsible
for which channel based on purely topographic evidence. Shifting of a river
mouth will, of course, immediately shift the location of the cross-shelf
terrigenous band and slow accumulation of carbonate sediment will become the
dominant process on the newly abandoned terrigenous wedge.

The expected resulting sediment column is depicted in cross section,
parallel to the shoreline, in Figure 8A. Each pod of river sediment grades
laterally into a thinner carbonate band. As a river mouth shifts, a
calcareous layer is deposited on top of the older river sand pod.
Presumably the total thickness of terrigenous sands would be at least ten
times that of the carbonate sequences, but representing an equivalent time
period.

Besides differing in thickness, calcareous and terrigenous layers also
differ in shape. The carbonate layers would be expected to be of uniform
thickness through the shelf area, whereas the terrigenous wedges are
variable in thickness and extent.

The difference in rates of sedimentation of the two sediment types should
produce topographic highs on the shelf where terrigenous material
predominates. Such shelf topography might in itself cause a shift of the
locus of deposition of fluvial sediments to adjacent lows or depressions
even without shifting of river mouths.

Off Laguna Tortuguero (Figs. 2 and 5), deposition of terrigenous sediment
appears to be topographically controlled through the following of
cross-shelf channels, assumed to be small river valleys cut during times of
lowered sea level. In such cases, the sediment column ultimately might
appear as shown in Figure 8B.

At least one example of an abandoned terrigenous sediment wedge is
located on the shelf north of Tortuguero Lagoon (Figure 5). At present, no
river drains onto this area of shelf. The wedge may be a remnant of sediment
once derived from the Cibuco River, which presently enters the sea just east
of the portion of the study area depicted in Figure 5. However the surficial
sediment tends to be somewhat more calcareous than that off other river
mouths indicating that carbonate sedimentation is just beginning to replace
terrigenous sedimentation.

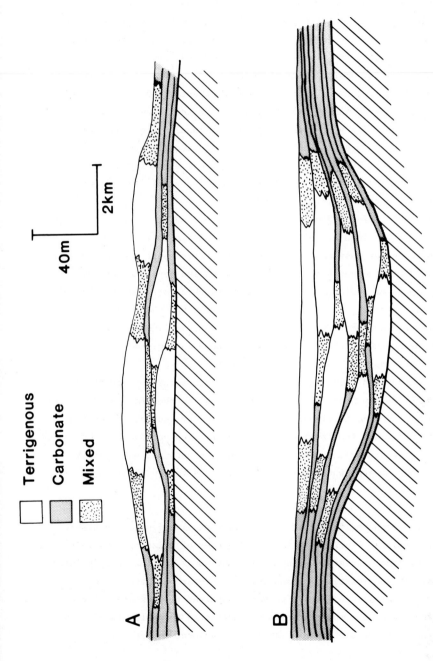

Fig. 8. Hypothetical shore-parallel cross sections of the shelf sediment cover of the north shelf assuming the absence of a nearby submarine canyon indenting the shelf. A) The sediment column on a flat, featureless shelf. Each terrigenous lobe represents a shift in river mouth location. B) The sediment column in a broad shelf channel fill.

The old Rio Cibuco shelf sediment pod differs from the active wedges of sediments off other river mouths in that it is quite thin (less than 10m thick) at the head of the submarine canyon (Figure 5). Off the Arecibo and Manati rivers, the greatest thicknesses of sediment accumulation, 30 to 40 meters, are found immediately adjacent to the canyon's head indenting the shelf. It is suggested that the portion of the Rio Cibuco sediment wedge adjacent to the canyon head has been removed by shelf processes since the river mouth shifted to the east. Resuspension of sediment during the frequent storms which characterize the north shelf of Puerto Rico, perhaps accompanied by mass movement processes, has apparently led to loss down the adjacent canyon system and ultimate deposition on the landward slope of the Puerto Rico trench. The very steep slope of the insular shelf adjacent to the submarine canyon systems indenting the shelf strongly support this process. The rims of the canyons of the north shelf typically extend to within 0.75-1.5km of the shoreline and to water depths of 30-40 meters. Slopes are of the order of 1:33 to 1:40.

The combination of the fine grain size of the fluvial contribution, the steep slopes of the insular shelf at the numerous deeply incised canyons and the high wave energy must continually combine to transport sediment seaward or canyonward off the north shelf. Under present sea level conditions, the presence of pods of sediment at canyon heads, such as off the Arecibo and Manati Rivers, probably can be maintained only through the continuing addition of new sediment. It seems likely that the terrigenous sediment accumulations on the north shelf have accumulated during and since the Holocene transgression.

It is hypothesized that, during a sea level transgression across the north shelf of Puetro Rico, the landward translating shoreface would largely remove any of the fine terrigenous sediment that had accumulated on the shelf during previous sea level highs. Again, a steep narrow shelf subjected to high wave energies would transport much of the unconsolidated shelf sediment cover seaward, whether or not a submarine canyon is nearby.

7 CONCLUSIONS

The narrow north shelf of Puerto Rico has a highly divergent sediment cover that consists of cross-shelf pods or wedges of relatively rapidly accumulating, fine terrigenous sands and muddy sands grading sharply into the more slowly accumulating, coarser, purely calcareous sands. The terrigenous wedges are restricted to areas adjacent to present or former river mouths; carbonate sedimentation is restricted to shelf areas between river mouths. The allochthonous terrigenous sediment wedges have accumulated in deposits as thick as 40 meters, whereas the coarser carbonate sand deposits are generally less than 10 meters thick. Outcrops of underlying Tertiary limestones frequently occur on the middle shelf in carbonate accumulation areas.

Due to shifting river mouths on the lower Puerto Rico coastal plain, the loci of deposition of the terrigenous sediment pods shift with time. Given a static sea level, the resultant sediment column should consist of laterally discontinuous terrigenous sands, tens of meters in thickness, separated by and laterally grading into much thinner calcareous layers.

At the head of submarine canyons, the north shelf is particularly narrow and steep and after a river mouth shift, terrigenous sand accumulations at canyon heads probably are rapidly lost down the canyon. Thus "permanent" accumulation of terrigenous sand bodies can occur only away from canyon heads.

It is also assumed that on this steep and narrow shelf, shoreline transgressions and regressions would cause removal of much of the accumulated terrigenous sand. Thus the present day accumulations of sediment on the north Puerto Rico shelf are largely a product of Holocene events. Preservation of insular shelf sediment columns such as that off the north shelf of Puerto Rico could be expected in conditions of active subsidence or long stillstands of high sea level.

ACKNOWLEDGEMENTS

This work was supported by the U. S. Geological Survey. We particularly wish to thank James V.A. Trumbull for his help in carrying out this study. Excellent cooperation from the Puerto Rico Department of Natural Resources made this study possible. Recent work on this study has also been supported by the National Science Foundation under grant #333-0721 to Pilkey.

REFERENCES

Adey, W.H., and Macintyre, I.G., 1973. Crustose coralline algae: A re-evaluation in the geological sciences: Bull. Geol. Soc. America., Vol. 84 p. 883-904.

Briggs, R.P., and Akers, J.P., 1965. Hydrogeologic map of Puerto Rico and adjacent islands: U.S. Geol. Survey Hydrologic Investigations, Atlas HA-197.

Department of Public Works, 1974. Oceanographic baseline data (1971-1972) for the formation of marine waste disposal alternatives of Puerto Rico: Vol. 1, Main Report; Vol. II, Appendices.

Ehlman, A.J., 1968. Clay mineralogy of weathered products and of river sediment, Puerto Rico: Jour. Sed. Petrology, v. 38, p. 885-889.

Fields, F.K., and Jordan, D.G., 1972. Storm-wave swash along the north coast of Puerto Rico: U.S. Geol. Survey, Hydrologic Investigations: Atlas HA-432, in two sheets.

Fuerst, S.I., 1979. Sediment transport on the insular slope of Puerto Rico off the la Plata River, unpublished MS Thesis, 101 p.

Grossman, Z.N., 1978. Distribution and dispersal of Manati river sediments: Puerto Rico Insular Shelf, unpublished MS Thesis, Duke University, Dept. of Geology, 72 p.

Grove, K.A., Pilkey, O.H., and Trumbull, J.V.A., 1982. Mud transportation on a steep shelf, Rio de la Plata Shelf, Puerto Rico: Geo-Marine Letters, v. 2, p. 71-75.

Lopez, M.A., and E. Colon-Dieppa, 1973. Magnitude and frequency of Floods in Puerto Rico: Co-operative Water Resource Investigation data release PR-9, 63 p.

Perkins, R.D., and Tsentas, C.I., 1976. Microbial infestation of Carbonate sub-strates planted on the St. Croix shelf, West Indies: Geol. Soc. Amer. Bull., p. 1615-1628.

Pilkey, O.H., Trumbull, J.V.A., and Bush, D.M., 1978. Equilibrium shelf sedimentation, Rio de la Plata Shelf, Puerto Rico: Jour. Sed. Petrology, v. 48, p. 389-400.

Pilkey, O.H., Fierman, E.I., and Trumbull, J.V.A., 1979. Relationship between physical condition of the carbonate fraction and sediment environments: Northern Puerto Rico shelf: Sedimentary Geology, v. 24, p. 283-290.

Pilkey, O.H., and Lincoln, R.B., 1984. Insular shelf heavy mineral partitioning: Northern Puerto Rico, Journal of Marine Mining, v. 4, p. 403-414.

Schneiderman, N., Pilkey, O.H., and Saunders, C., 1976. Sedimentation on the Puerto Rico insular shelf: Jour. Sed. Petrology, v. 46, p. 167-173.

Seilacher, A., 1982. Distinctive features of sandy tempestites: In Einsele, G. and Seilacher, A. (eds.): Cyclic and event stratification: 333-349, Berlin, Heidelberg, New York (Springer).

Shepard, F.P., Marshall, N.F., McLoughlin, P.A., and Sullivan, G.G., 1979. Currents in submarine canyons and other seavalleys: Amer. Assoc. Petroleum Geologists, Studies in Geology No. 8: Tulsa, OK, 173 p.

Stout, P.M., 1979. Calcium carbonate sedimentation on the northeast insular shelf of Puerto Rico, unpublished MS Thesis, Duke University, Department of Geology, 107 p.

Swift, D.J.P., 1976. Coastal sedimentation, in Stanley, D.J., and Swift, D.J.P., eds., Marine Sediment Transport and Environmental Management: New York, Jolhn Wiley and Sons, p. 255-310.

Western Geophysical Company of America, 1975. Shallow water bathymetric, sonar, and seismic investigations for siting the north-coast nuclear plant (NORCO-NP-1) on Puerto Rico, Appendix 2.5R, Amendment 27.

Wood, E.D., Youngbluth, J.J., Nutt, M.E., Yeaman, M.N., Yoshioku, P., and Canoy, M.J., 1975, Punta Manati Environmental Studies: Puerto Rico nuclear center, Publication #182, 225p.

Chapter 9

ACCUMULATION OF MIXED CARBONATE AND SILICICLASTIC MUDS ON THE CONTINENTAL
SHELF OF EASTERN SPAIN

C.A. NITTROUER, B.E. BERGENBACK, D.J. DeMASTER and S.A. KUEHL
Department of Marine, Earth and Atmospheric Sciences, North Carolina State
University, Raleigh, North Carolina 27695

ABSTRACT
        The continental shelf of eastern Spain (extending from the Ebro
River southward through the Gulf of Valencia) is accumulating a mixture of
carbonate and siliciclastic sediment. Twenty-eight cores were analyzed to
characterize sediment size and composition and to measure mixing and
accumulation rates. Sandy mud consisting of about 40% carbonate and 60%
siliciclastic sediment is accumulating over a transgressive sand and gravel
layer, which is exposed landward of the 20 m isobath. Profiles of Pb-210
indicate that much of the mud deposit has an apparent accumulation rate of
about 1 mm/yr. Profiles of Cs-137, however, indicate that in some cores the
Pb-210 rates overestimate the true accumulation rates. The observed
accumulation is slow compared to dispersal systems associated with larger
fluvial sources of siliciclastic sediment. The slow accumulation combined
with moderate sediment mixing rates (determined from Th-234 profiles)
predict preservation of homogeneous strata, which is observed in
x-radiographs. Although the siliciclastic accumulation is relatively slow,
the carbonate accumulation rates are similar to values for low-latitude
carbonate environments. The slow accumulation of siliciclastic sediment
combined with apparently high productivity of carbonate sediment results in
the mixed carbonate and siliciclastic deposit. The special characteristics
of the study area are: a) siliciclastic sediment flux to the shelf is small
but significant; b) the physical-oceanographic regime is quiescent enough to
allow much of the fine sediment to accumulate on the shelf; c) and shelf
waters are warm (for mid-latitudes) allowing rapid production of carbonate
sediment.

1 INTRODUCTION

        The topic of shallow carbonate sedimentation contains several

well-established principles (Wilson, 1975), including: carbonate sediments

are restricted to low latitudes (<30°); carbonates are excluded from areas

of fine grained siliciclastic input; and significant deposits of carbonate

are nearly pure carbonate. These principles are largely derived from studies

of reef systems, which are probably the most-interesting carbonate

depositional environments. Similarly, the topic of active sedimentation on

continental shelves generally has emphasized siliciclastic components to the

exclusion of other sediment types. This is because the most-interesting

accretionary shelves are associated with dispersal systems of large rivers

(annual sediment discharge $>10^7$ metric tons), where siliciclastic sediment

overwhelms other material. However, not all carbonate is associated with
reef systems, and not all accretionary shelf deposits are associated with
large rivers.

The present study investigates a continental shelf region where a
mixture of carbonate and siliciclastic sediment is accumulating. The
objectives are to characterize the sediment (grain size and composition) and
to measure dynamic aspects of strata formation (mixing rate and accumulation
rate), in order to contrast the local sedimentation with that observed for
regions of primarily carbonate and primarily siliciclastic sedimentation.

2 BACKGROUND

2.1 Regional setting

The Ebro River is the primary source of siliciclastic sediment for
the study area (Fig. 1), which encompasses the continental shelf from the
river mouth southward through the Gulf of Valencia to 39°N. The drainage
basin for the Ebro River represents approximately 15% of the Iberian
Peninsula and includes the southern slopes of the Pyrenees Mountains. A
diverse suite of rocks is found within the drainage basin, ranging in age
from Paleozoic to Cenozoic and including igneous, metamorphic and
sedimentary types. The continental margin underwent subsidence during the
Tertiary and accumulated up to 4000 m of sediment (Maldonado, 1975). The
margin is stable today, although Pleistocene deposits are significant in
some locations and Holocene deposits are relatively thick near the mouth of
the Ebro. The delta at the mouth of the Ebro attests to the quiescence of
the local marine environment and to the potential of the river as a supplier
of sediment to the adjacent continental shelf. Other rivers entering the
study area drain arid regions and their limited discharge of water and
sediment is eliminated by dams constructed for agricultural purposes. The
annual suspended-sediment discharge of the Ebro River is approximately 2.5 x
10^6 tons (Maldonado, 1972).

Within the study area (Fig. 1), the shelf break occurs at about
160 m water depth (Maldonado, 1975), and the distance from shore to the
break decreases from 75 km near the mouth of the Ebro River to 25 km at
39°N. The effects of tides and waves on sedimentation in the study area are
relatively unimportant. Tides are semi-diurnal with a maximum range of only
20 cm (Maldonado, 1972). Waves on the western side of the Mediterranean are
less than one meter in height 90% of the year, and wavelengths are less than
forty meters 80% of the year (Jago and Barusseau, 1981). Currents on the
shelf respond to seasonal winds (Maldonado et al., 1983). Southeasterly
winds, which blow during the summer months, create gentle northeastward

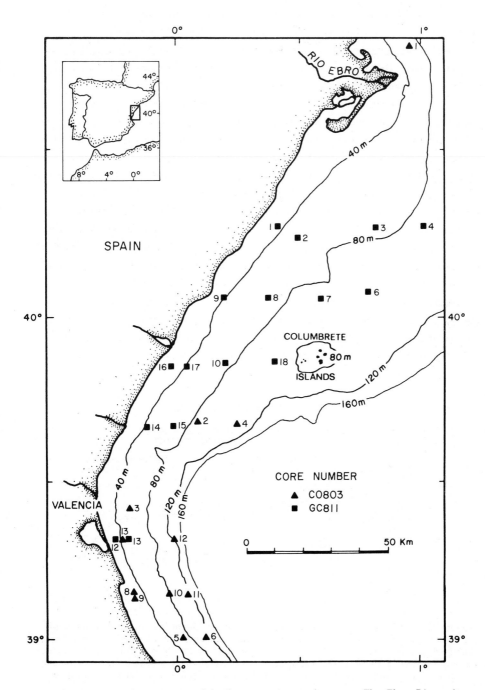

Figure 1. Box core locations and bathymetry in study area. The Ebro River is the dominant source of siliciclastic sediment, and transport is southward.

currents. Northeasterly winds (called levants), originating in the Alps, bring storms between November and March. The levants blow toward the southwest (parallel to the coast) and create a barotropic flow of water to the southwest. The associated storms are believed to generate the greatest water velocities and to transport the largest amount of sediment.

Mud covers the shelf, except for the area shoreward of the 20 m isobath, where sand and gravel are present. The mud includes both siliciclastic and carbonate components. The carbonate component primarily is modern biogenic debris, with the dominant source changing in an offshore direction from mollusks (inner shelf) to benthic foraminifera (middle shelf) to pelagic foraminifera (outer shelf) (Maldonado et al., 1983). General models of continental shelf sedimentation (Curray, 1965; Swift, 1970) predict that modern mud should be deposited upon a basal transgressive sand and gravel layer. In these models, the transgressive layer is the product of shoreface erosion associated with the Holocene Transgression. Fine-grained sediment deposited in shallow water is resuspended and ultimately accumulates below wave-base. In the study area, mud covers much of the shelf, because of the quiescent wave regime. The stratigraphy of the shelf in the study area is depicted in Figure 2. Geophysical studies demonstrate that the surficial mud layer is generally 1-10m thick, but in some places

VALENCIA SECTION

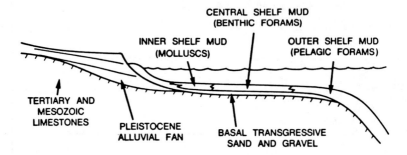

Figure 2. Schemetatic cross-section of Holocene deposits in the study area (Maldonado et al., 1983). A transgressive sand and gravel layer is exposed landward of the 20m isobath, and is covered by a sandy mud deposit in deeper water. The mud contains 40% carbonate material, whose predominant component changes in a seaward direction from molluscs to benthic forams to pelagic forams.

is reduced to negligible thicknesses due to antecedent topography (Maldonado et al., 1983; Rey and Diaz Del Rio, 1983).

2.2 Geochronology

The concepts of geochronology, as they are applied to this study, are summarized in two recent papers: Nittrouer et al. (1984a) and DeMaster et al. (1985). The relevant conclusions of these papers are briefly presented here, and more thorough discussions can be found in the papers themselves. Steady-state profiles of Th-234 (half-life 24 days) and Pb-210 (half-life 22.3 years) in the seabed are generally controlled by sediment mixing and accumulation, which are characterized, respectively, by a mixing coefficient (D, in cm^2/yr) and an accumulation rate (A, in cm/yr). A and D are related by the following equation:

$$A = \frac{\lambda z}{\ln(C_o/C_z)} - \frac{D}{z} (\ln C_o/C_z) , \qquad (1)$$

where λ is the decay constant (in yr^{-1}) for a particular radioisotope, C_o is its radioactivity (in disintegrations per minute per gram) in the upper portion of a profile, and C_z is its radioactivity a distance z deeper in the seabed. Biological mixing usually dominates the upper 5-10 cm of the seabed, and for this region Equation 1 can be simplified to:

$$D = \lambda \left(\frac{z}{\ln(C_o/C_z)}\right)^2. \qquad (2)$$

A profile of excess Th-234 (i.e., radioactivities in excess of those supported by its parent U-238) and Equation 2 commonly are used to evaluate sediment mixing. Below the surface mixed layer, mixing usually is assumed to be negligible, and for this region Equation 1 can be simplified to:

$$A = \frac{\lambda z}{\ln(C_o/C_z)} . \qquad (3)$$

A profile of excess Pb-210 (i.e., radioactivities in excess of those supported by its effective parent Ra-226) and Equation 3 commonly are used to evaluate sediment accumulation. Comparison of Equations 1 and 3 demonstrates that if the assumption of no deep mixing is wrong, then Equation 3 will overestimate the true accumulation rate. The assumption can be tested from the penetration depth within the seabed of the bomb-produced radioisotope Cs-137, which has been present in the marine environment for about 30 years (since the early 1950's). If particles are mixed rapidly throughout the surface layer, Cs-137 should be found below this layer no deeper than the product of 30 years and the Pb-210 annual accumulation rate. If Cs-137 is found deeper, the assumption of no deep mixing is wrong, and the apparent Pb-210 accumulation rate (from Equation 3) is a maximum estimate.

3 METHODS

3.1 Field sampling

Twenty-eight cores were obtained from the study area (Fig. 1) during cruises completed in May-June 1980 (CO803) and February 1981 (GC811). Cruise CO803 collected large-diameter cores (15cm x 15cm cross section) and GC811 collected box cores (20cm x 30cm cross section); core lengths ranged from 11 to 78 cm. One-centimeter samples were obtained at numerous depths for Pb-210 analysis in most cores and for Th-234 analysis in two cores. Large-interval samples (up to 5 cm) were taken from most cores for Cs-137, geochemical, and sedimentological analyses. In addition, x-radiographs (3 cm thick) were obtained for many cores.

3.2 Sedimentological analyses

Grain size was determined by standard sieve and pipette techniques (Krumbein and Pettijohn, 1938). Measurements were completed at $1/4$ ϕ intervals in the sand fraction, at $1/2$ ϕ intervals in the silt fraction, and at 1 ϕ intervals in the clay fraction.

3.3 Geochemical analyses

Calcium carbonate content was determined by an acid-treatment weight-loss technique (Gross, 1971). Dried sediment was treated with a dilute hydrochloric-acid solution (2.5 N) and the residue was collected on a 0.45 μm filter. After drying the filter, the observed weight loss was converted into percent calcium carbonate. Surface samples were analyzed in each core. Three cores were examined for down-core trends in carbonate content. In three additional cores, the relationship between carbonate content and grain size was examined. Sediment was separated by sieving into a sand fraction (coarser than 64 μm) and a silt-clay fraction (finer than 64 μm).

Organic-matter content was determined by the wet-chemical technique described by Gaudette et al. (1974). Biogenic-silica content was analyzed using the sodium-carbonate leaching technique of DeMaster (1981).

3.4 Radiochemical analyses

Th-234 was measured in two cores (TC-2 and TC-13) from cruise GC811. Th-234 analyses followed the technique described by Aller and Cochran (1976). Pb-210 analyses were completed using the Pb-210 technique described by Nittrouer et al. (1979). In order to directly determine the supported (or background) level of Pb-210 activity, Ra-226 was measured in cores by the Rn-222 emanation technique (see Benninger, 1976). Cs-137 was measured by gamma-ray emission of sediment samples using a germanium detector at North Carolina State University and using a GeLi detector at Oak Ridge National Laboratory. Sediment samples were counted in plastic petri dishes (125 cm^3) or Marinelli beakers (550 cm^3).

4 RESULTS

4.1 Grain size

Most cores were obtained seaward of the nearshore deposit of sand and gravel. Therefore, most of the observed sediment is fine-grained (i.e., >50% silt and clay). Near the mouth of the Ebro River and in the Gulf of Valencia median grain sizes range from 5ϕ to 8ϕ (Fig. 3). Between these two areas, sediment is coarser, and median sizes range from 3ϕ to 5ϕ. Significant vertical trends are not observed in cores (Bergenback, 1984).

4.2 Sediment composition

The calcium-carbonate content of surface sediment is uniformly about 40% (Fig. 4). Significant trends are not observed with distance from the Ebro River nor with distance from shore. Carbonate content shows no significant change with depth in the seabed (Bergenback, 1984). Analyses of the sand and mud fractions reveal that carbonate can be associated with either. Carbonate content in the sand fraction ranges from about 40% to 70% and in the silt-clay fraction ranges from about 30% to 60%. Replicate analyses demonstrate that the precision of the carbonate analysis is ± 2%. The content of organic matter and of biogenic silica are negligible (<1 %).

4.3 Mixing coefficients and accumulation rates

The two cores examined for Th-234 reveal that excess activities extend to a depth of 4-5 cm (Fig. 5). Mixing coefficients calculated from the slopes of the profiles (Equation 2) are 29 and 56 cm^2/yr for TC-2 and TC-13, respectively.

258

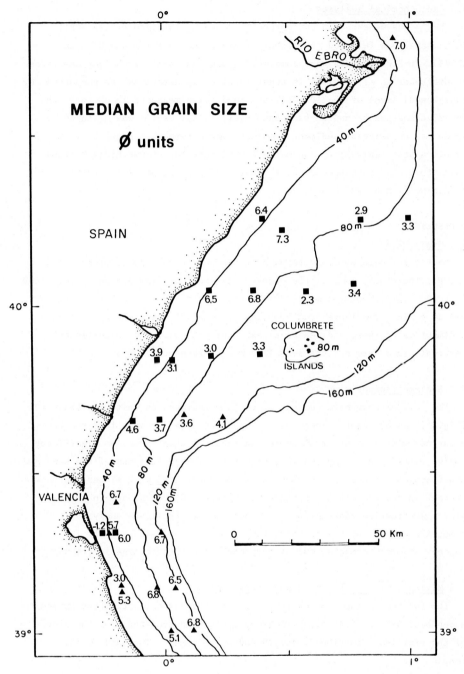

Figure 3. Median grain size in surface sediment of box cores. Near the Ebro
River and in the Gulf of Valencia median sizes range from 5φ to 8φ.
Extending across the central portion of the shelf is a coarser region,
where median sizes range from 3φ to 5φ.

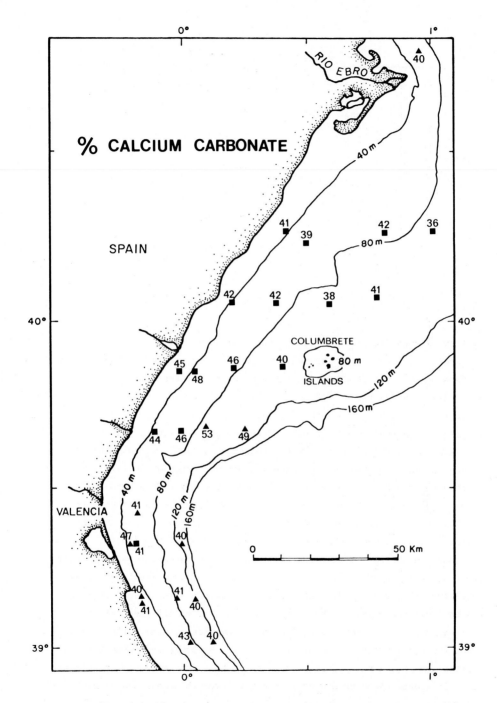

Figure 4. Carbonate content in surface sediment (analytical precision ± 2%).
Significant vertical changes are not observed in cores. Carbonate is
present in both the sand and mud fractions.

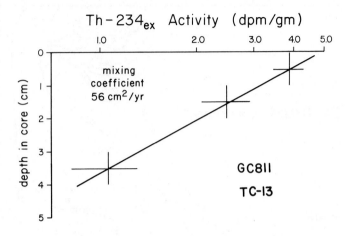

Figure 5. Profile of excess Th-234 at TC-13 (GC811), indicating a mixing coefficient of 56 cm^2/yr.

A typical Pb-210 profile for the study area is shown in Figure 6. For this core, uniform activities near the surface indicate a thickness for the surface mixed layer of about 4 cm. For other cores, the thickness of the mixed layer ranges fro 0 to 5 cm. Below the surface layer, Pb-210 activity decreases logarithmically downward, reaching a supported level at about 1 dpm/gm. Supported levels in other cores range from 0.6 to 1.1 dpm/gm, and are verified by Ra-226 measurements (Bergenback, 1984). By assuming that no mixing occurs below the surface layer, apparent accumulation rates can be calculated from the slope of the excess Pb-210 profile below the surface layer (Equation 3). These rates decrease from 2.9 mm/yr near the mouth of the Ebro River to <1 mm/yr south of the delta (Fig. 7). In the Gulf of Valencia, apparent accumulation rates range from 0.6-1.9 mm/yr.

Using the observed mixing coefficients and mixing depths together with the Dispersion Equation (see Nittrouer et al., 1984a) demonstrates that particles are mixed rapidly throughout the surface layer (i.e., on time scales less than a year). Therefore, the depth of Cs-137 penetration into the seabed is useful for evaluating the assumption of negligible mixing below the surface layer (see Background). In the study area, the Cs-137 penetration is generally less than 10 cm, but in some cases the penetration depth is greater than predicted from mixed layer thickness and Pb-210 accumulation rate. For example, at TC-13 (Fig. 6) the predicted depth of

261

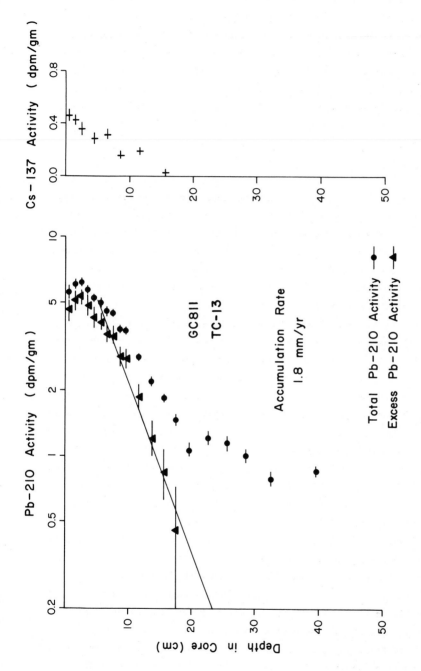

Fig. 6. Profiles of Pb-210 and Cs-137 at TC-13 (GC811). Pb-210 activities near the surface indicate a mixed layer about 4 cm thick. Below this surface mixed layer, the logarithmic decrease in excess Pb-210 indicates an accumulation rate of 1.88mm/yr. The depth of Cs-317 penetration into the seabed suggests that this is a maximum estimate of accumulation rate (see text).

Cs-137 penetration is 9-10 cm, but the observed depth is at least 11-12 cm
(the data point at 15-16 is indistinguishable from 0). Therefore, the Pb-210
accumulation rate overestimates the true accumulation rate at this station.
The degree of error is difficult to evaluate because Pb-210 profiles are
very sensitive to mixing, if the accumulation rate is low (~1mm/yr) (see
DeMaster et al., 1984). The stations which showed discrepancies between
observed and predicted depths of Cs-137 penetration are noted in Figure 7,
and accumulation rates indicated for these stations should be considered
maximum estimates.

5 DISCUSSION

5.1 Comparison with siliciclastic sedimentation on other continental shelves

In the most general sense, sedimentation in the study area resembles
siliciclastic sedimentation associated with other fluvial dispersal systems.
This is demonstrated by the presence of a Holocene mud deposit overlying a
transgressive sand and gravel layer. However, several aspects of the
regional setting impart unique character to the mud deposit. A region of
relatively coarse sediment and slow accumulation extends across the central
portion of the study area. These characteristics might result from
significant thinning of the mud deposit due to antecedent topography and/or
from intensification of flow due to the abrupt narrowing of shelf width. In
either case, the presence of excess Pb-210 and of Cs-137 in the seabed
indicates that new sediment is reaching the central portion of the study
area, but that net accumulation is slow or negligible. Therefore, sediment
is largely bypassing this region.

Probably the most important characteristic of siliciclastic sedimentation
in the study area is the low sediment supply. The sediment discharge of the
Ebro River (~10^6 tons/year) is significantly smaller than that of other
rivers where accretion of shelf sediment has been well documented, for
example: Columbia River (~10^7 tons/year), Yangtze River (~10^8 tons/year),
and Amazon River (~10^9 tons/year). The accumulation rates on the adjacent
continental shelves are similarly greater than in the present study area;
several times greater for the shelf associated with the Columbia River
(Nittrouer et al., 1979) and at least an order of magnitude greater for the
shelves associated with the Yangtze and Amazon Rivers (DeMaster et al.,
1984; Kuehl et al., 1981). The diminutive nature of siliciclastic
accumulation in the study area is further emphasized by recognition that 40%
of the accumulating sediment is biogenic carbonate and that negligible

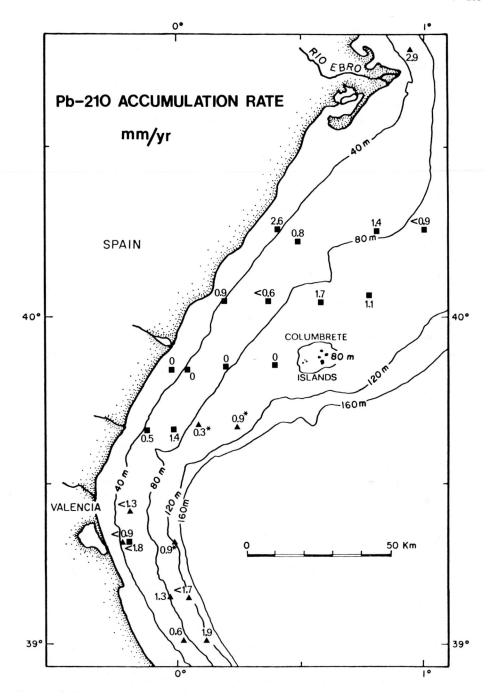

Figure 7. The distribution of accumulation rates in the study area. The stations are indicated with the symbol < where rates are maximum estimates and with the symbol * where Cs-137 data are not available. A region of low or negligible accumulation rate extends across the central portion of the study area. Otherwise values are about 1 to 2 mm/yr.

amounts of carbonate are accumulating in the other shelf systems. However, the relationship between low siliciclastic flux and high carbonate flux is probably not coincidental, as will be discussed in the next section.

Establishment of a sediment budget from the present data base is difficult, because of the limited cores near the mouth of the Ebro River and because of the ambiguity associated with some accumulation rates. However, calculations suggest that at least 1.7×10^6 tons of siliciclastic sediment annually accumulate on the shelf (Bergenback, 1984). This represents about two-thirds of the discharge from the Ebro River, and would be much smaller if the physical regime were more typical of oceanic conditions and capable of greater rates of sediment erosion and transport. If the study area were situated along the coast of a major ocean (e.g., East Coast U.S.) instead of an enclosed sea, the mud probably would be the basal sand and gravel layer.

The nature of sedimentary structure preserved by strata formation is dependent on the ratio of mixing rate to accumulation rate (Guinasso and Schink, 1975; Nittrouer and Sternberg, 1981). The mixing coefficients for the study area are typical of continental shelves, but the accumulation rates are relatively low. This results in ratios of mixing rate to accumulation rate of about 90 and 80 for TC-2 and TC-13, respectively (the value for TC-13 should be even higher, because the accumulation rate is a maximum estimate). Values >10 indicate destruction of primary structure and preservation of homogeneous strata. This character can be seen in the x-radiograph for TC-13 (Fig. 8), and represents an important result of the low accumulation rates. High ratios of mixing rate to accumulation rate are commonly present in distal portions of large dispersal systems (Nittrouer et al., 1984b). For the present study area, high ratios characterize the entire dispersal system.

5.2 Comparison with Carbonate Sedimentation in Other Shallow Environments

Accumulation rates about and above 1 mm/yr are common in the present study area, and carbonate represents 40% of this accumulation. Although such rates are small compared to siliciclastic accretionary settings, the rates are similar to rapidly accumulating carbonate settings. Wilson (1975) tabulates Holocene accumulation rates for a variety of low-latitude locations, and the average rate for shallow environments is about 1 mm/yr. Although Wilson's definition of shallow is more restrictive (lagoon, tidal flat, sabkha, reef), the present study uses shallow to refer to marine environments landward of the shelf break. Considering the present study area, appreciable shallow carbonate accumulation can occur outside low latitudes.

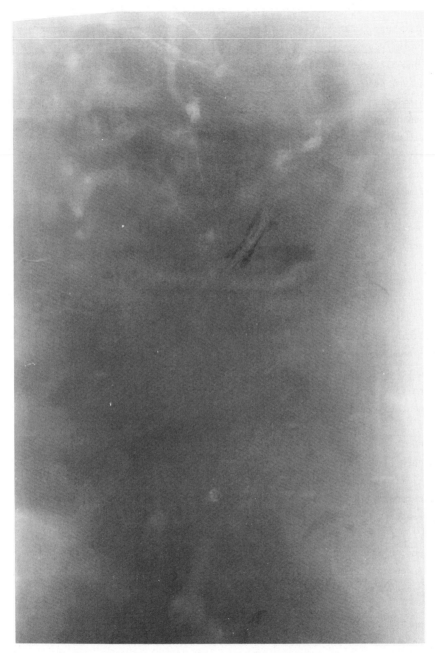

Figure 8. X-radiograph (positive) of box core at TC-13 (GC811), dimensions
11cm x 35 cm. This core contains relatively homogeneous strata with no
evidence of primary sedimentary structure. The profiles of Th-234 and
Pb-210 at this station (Figs. 5 and 6) indicate a ratio of mixing rate
to accumulation rate of at least 80. Ratios greater than 10 predict
homogeneous structure.

Bioclastic carbonate deposits are recognized in a number of mid- and
high-latitude locations (e.g.: Alaska, Hoskin and Nelson, 1969; southern
Australia, Wass et al., 1970; New Zealand, Nelson et al., 1982, Portugal,
Dias and Nittrouer, 1984). However, these are nearly pure carbonate, sand
and gravel deposits. They form due to strong physical winnowing and little
or no dilution from siliciclastic sediment. Typical carbonate concentrations
on continental shelves are 4-6% (Milliman, 1974) and even these values are
restricted to relict or palimpsest deposits (e.g., East Coast U.S.). The
rate of siliciclastic input is generally recognized to be the primary
control for concentration of carbonate in mid- and high-latitude deposits
(Milliman, 1974; Leonard et al., 1981). The input of siliciclastic sediment
not only dilutes carbonate sediment, but high fluxes of siliciclastic
sediment cause deterioration of living conditions (especially for suspension
feeders) and reduction in productivity. However, considering the present
study area, appreciable carbonate accumulation can occur under conditions of
small (but significant) input of fine-grained siliciclastic sediment.

The secondary control on carbonate concentration in shallow environments
is the rate of productivity, which is largely dependent on water temperature
(Milliman, 1974; Leonard et al., 1981). The solubility of $CaCO_3$ decreases as
water temperature increases, and carbonate organisms are generally more
productive in warm water. The western Mediterranean Sea is a mid-latitude
region (35°-45°) which contains many areas rich in shallow carbonate
sediment (e.g., Alborean Sea, Milliman et al., 1971; Algerian Shelf, Caulet,
1971). These are nearly-pure (>90%) carbonate deposits, which are partly
dependent on a lack of significant siliciclastic input. However, along the
southern (Zamerreno et al., 1983) and the eastern (this study) shelves of
Spain, mixed carbonate and siliciclastic deposits are present and reflect
the secondary importance of high productivity. In this regard the water
temperatures (24°C summer, 14°C winter; Julia, 1983), which are relatively
warm for the latitude, are probably responsible for allowing rapid
production of carbonate sediment. The present study area demonstrates that
under conditions of high carbonate productivity and small (but significant)
siliciclastic input, a mixed carbonate-siliciclastic deposit can form.

6 CONCLUSIONS

1) Sandy muds are accumulating at rates of about 1 mm/yr over much of the
   study area. These sediments are about 40% carbonate material from
   skeletal remains and 60% siliciclastic material from the Ebro River.

2) The supply and accumulation of siliciclastic sediment in the study area is small compared to other fluvial dispersal systems. That fine sediment accumulates at all is probably the result of a physical-oceanographic regime which is quiescent relative to typical oceanic settings. Low accumulation rates combined with moderate mixing rates produce homogeneous strata, such as found in distal portions of larger dispersal systems.

3) Carbonate sediment is accumulating in this mid-latitude location at rates similar to low-latitude carbonate environments. The small (but significant) input of fine-grained siliciclastic sediment neither eliminates carbonate production by organisms nor overwhelms accumulation of carbonate sediment. The result is a mixed carbonate-siliciclastic deposit.

7 ACKNOWLEDGEMENTS

This research was part of a joint Spanish-American cooperative Program in Oceanography. Funding for research at North Carolina State University came from the Atlantic Oceanographic and Meteorological Laboratories of NOAA. D.J.P. Swift, R.A. Young, and A. Maldonado were responsible for most of the scientific and logistical coordination of the program. They also were responsible for involvement of scientists from NCSU and for much of the success of the present paper. The authors thank N.H. Cutshall and I.L. Larsen for making some of the Cs-137 measurements. This paper is contribution no. 8500, Department of Marine, Earth and Atmospheric Sciences, North Carolina State University.

8 REFERENCES

Aller, R.C., and Cochran, J.R., 1976. ^{234}Th - ^{238}U disequilibrium and diagenetic time scales: Earth Planet. Sci. Letters, v. 29, p. 37-50.

Benninger, L.K., 1976. The uranium-series radionuclides as tracers of geochemical processes in Long Island Sound [unpub. Ph.D. thesis]: New Haven, Yale University, 151 p.

Bergenback, B.E., 1984. Modern sediment accumulation on the eastern continental shelf of Spain, south of the Ebro River [unpub. M.S. thesis]: Raleigh, North Carolina State University.

Caulet, J.P., 1971. Recent biogenic calcareous sedimentation on the Algerian continental shelf, in Stanley, D.J., ed., The Mediterranean Sea: a Natural Sedimentation Laboratory: Stroudsburg, Dowden, Hutchinson and Ross, p. 261-277.

Curray, J.R., 1965. Late Quaternary history, continental shelves of the United States, in Wright, H.E., and Frey, D.C., eds., The Quaternary of the United States: Princeton Univ. Press, p. 723-735.

DeMaster, D.J., 1981. The supply and accumulation of silica in the marine environment: Geochim. Cosmochim. Acta, v. 45, p. 1715-1732.

DeMaster, D.J., McKee, B.A., Nittrouer, C.A., Qian, J., and Cheng, G., 1985. Rates of sediment accumulation and particle reworking based on radiochemical measurements from continental shelf deposits in the east China Sea: Cont. Shelf Res., in press.

Dias, J.M.A. and Nittrouer, C.A., 1984. Continental shelf sediments of northern Portugal: Cont. Shelf Res., v. 3, p. 147-165.

Gaudette, H.E., Flight, W.R., Toner, L., and Folger, D.W., 1974. An inexpensive titration method for the determination of organic carbon in recent sediments: Jour. Sed. Petrol., v. 44, p. 249-253.

Gross, M.G., 1971. Carbon determination, in Carver, R.E., ed., Procedures in sedimentary petrology: New York, John Wiley and Sons, p. 573-596.

Guinasso, N.L., and Schink, D.R., 1975. Quantitative estimates of biological mixing rates in abyssal sediments: J. Geophys. Res., v. 80, p. 3032-3043.

Hoskin, C.M., and Nelson, R.V., 1969. Modern marine carbonate sediment, Alexander Archipelago, Alaska: Jour. Sed. Petrol., v. 39, p. 581-590.

Jago, C.F., and Barusseau, J.P., 1981. Sediment entrainment on a wave-graded shelf, Roussillon, France: Mar. Geol., 4. 42, p. 279-299.

Julia, A., 1983. Utilizacion de las series temporales de temperatura y salinidad para interpretar los movimientos de las masas de aqua (zona de Amposta), in Castellvi, J., ed., Estudio Oceanografico de la Plataforma Continental: Proc. Spanish-American Oceanographic Meeting, Cadiz, p. 163-171.

Kuehl, S.A., Nittrouer, C.A., and DeMaster, D.J., 1982. Modern sediment accumulation and strata formation on the Amazon continental shelf: Mar. Geol., v. 49, p. 279-300.

Krumbein, W.C., and Pettijohn, F.J., 1938. Manual of Sedimentary Petrography: New York, Appleton-Century-Crofts, 549 p.

Leonard, J.E., Cameron, B., Pilkey, O.H., and Friedman, G.M., 1981. Evaluation of cold-water carbonates as a possible paleoclimatic indicator: Sed. Geol., v. 28, p. 1-28.

Maldonado, A., 1972. El delta del Ebro [unpub. Ph.D. thesis]: Barcelona, Univ., Barcelona, 486 p.

Maldonado, A., 1975. Deltas of the northern Mediterranean Sea: the Ebro Delta, Field guide to trip 16, IXth Int. Congr. Sed., Nice, 78 p.

Maldonado, A., Swift, D.J.P., Young, R.A., Han, G., Nittrouer, C.A., DeMaster, D.J., Rey, J., Palomo, C., Acosta, J., Ballester, A., and Castellvi, J., 1983, Sedimentation on the Valencia continental shelf: preliminary results: Cont. Shelf Res., v. 2, 195-211.

Milliman, J.D., 1974. Marine Carbonates: New York, Springer-Verlag, 375 p.

Milliman, J.D., Weiler, Y., and Stanley, D.J., 1971. Morphology and carbonate sedimentation on shallow banks in the Alborean Sea, in Stanley, D.J., ed., the Mediterranean Sea: A Natural Sedimentation Laboratory: Stroudsburg, Dowden, Hutchinson and Ross, p. 241-259.

Nelson, C.S., Hancock, G.E., and Kamp, P.J.J., 1982. Shelf to basin, temperate skeletal carbonate sediments, Three Kings Plateau, New Zealand: Jour. Sed. Petrol., v. 52, 717-732.

Nittrouer, C.A., Sternberg, R.W., Carpenter, R., and Bennett, J.T., 1979. The use of Pb-210 geochronology as a sedimentological tool: application to the Washington continental shelf: Mar. Geol., v. 31, p. 297-316.

Nittrouer, C.A., and Sternberg, R.W., 1981. The formation of sedimentary strata in an allochthonous shelf environment: the Washington continental shelf: Mar. Geol., v. 42, p. 201-232.

Nittrouer, C.A., DeMaster, D.J., McKee, B.A., Cutshall, N.H., and Larsen, I.L. 1984a. The effect of sediment mixing on Pb-210 accumulation rates for the Washington continental shelf: Mar. Geol., v. 54, p. 201-221.

Nittrouer, C.A., DeMaster, D.J., and McKee, B.A., 1984b. Fine-scale stratigraphy in proximal and distal deposits of sediment dispersal systems in the East China Sea: Mar. Geol., v. 61, p. 13-24.

Rey, J., and Diaz Del Rio, V., 1983. Aspectos geologicos, sobre la estructura poco profunda de la plataforma continental del levante espanol, in Castellvi, J., ed., Estudio Oceanografico de la Plataforma Continental: Proc. Spanish-American Oceanographic Meeting, Cadiz, p. 53-74.

Swift, D.J.P., 1970. Quaternary shelves and the return to grade: Mar. Geol., v. 8, p. 5-30.

Wass, R.E., Conolly, J.R., and MacIntyre, R.J., 1970. Bryozoan carbonate sand continuous along southern Australia: Mar. Geol., v. 9, p. 63-73.

Wilson, J.L., 1975. Carbonate Facies in Geologic History: New York, Spring-Verlag, 471 p.

Zamerreno, I., Vazquez, A., and Maldonado, A., 1983. Sedimentacion en la plataforma de Almeria: un ejemplo de sedimentacion mixta silico-carbonatada en clima templado, in Castellvi, J., ed., Estudio Oceanografico de la Plataforma Continental: Proc. Spanish-American Oceanographic Meeting, Cadiz, p. 97-119.

Chapter 10

CARBONATE TO SILICICLASTIC PERIPLATFORM SEDIMENTS: SOUTHWEST FLORIDA

C.W. HOLMES Department of Interior P.O. Box 6732, Corpus Christi, Texas 78411

ABSTRACT
    Three distinct carbonate deposits have been identified on the slope and
adjacent sea floor of the southwestern Florida Platform: (1) reef talus,
recognized by shape and location, found on the upper slope of the Yucatan
Channel and also east of the Marquesas Keys; (2) hemipelagic sediments, with
complex sigmoid-oblique bed forms, filling the intervening gap between the
channel and Keys and forming two lobes on the floor of the northern Florida
Straits; and (3) turbidite deposits, with chaotic internal bed forms, covering
siliciclastic Mississippi Fan sediments at the base of the canyons in the
Florida escarpment. The source of the talus, eroded and transported during the
many storms that frequent the region, is the reef complexes that have formed on
the platform rim. The sediment of the other two deposits is of foraminiferal
tests, produced in nutrient-rich waters at the shelf edge. This sediment is
deposited on the outer shelf and is vigorously transported southward, as
evidenced by 5 m high asymmetric sand waves.
    Geophysical, geochemical, and sedimentological data suggest that the
spatial relationships of these deposits are related to sea-level variations.
During extreme lowstands, with much of the shelf exposed, the dominant
sedimentation was in the form of siliciclastic deposition on the abyssal floor,
and slope talus development at the edge of the shelf. During a subsequent rise
in sea level, after carbonate production on the shelf was initiated, sediment
was transported southward to the head of the canyons and funneled to the
abyssal floor. Subsequent rising sea level shifted the axis of transport
farther to the shelf, bypassing the canyons and funneling the sediment through
breaks in the carbonate reef banks at the southern edge of the platform. At
the sites of both the hemipelagic and the turbidite deposition, high-resolution
seismic data indicate that at least three cycles of deposition have occurred.
In the abyss, this cyclic nature has produced alternating layers of carbonate
and noncarbonate sediments, recognizable in the sedimentary record as limestone
units interlayered with fine shales. In the geologic record the hemipelagic
deposits would be almost indistinguishable from deep-sea foraminiferal oozes.

1 INTRODUCTION

    Owing to their different origin, sequences of intercalated carbonate and

noncarbonate rocks pose an intriguing geologic problem. Carbonates deposited in

any volume are dominantly biogenic and marine in origin. Because of their mode

of formation, they are moderately fragile and are unable to withstand extensive

turbulence or transportation. Noncarbonate sediments, the majority of which are

composed of silica-based minerals, are products of weathering in exposed

terranes. These sediments, predominantly silts and clays, are often transported

significant distances before being deposited and are able to withstand

significant turbulence without much alteration. In general, because these

sediments do not suffer biogenic degradation, net accumulation rates are

greater than those of carbonates. Thus, to find strata of alternating composition, a somewhat unique environmental depositional setting must be present.

The present volume has many examples of coexisting carbonate/noncarbonate
sediments. Choi and Holmes (1983) presented evidence on the formation of
carbonate reefs on riverine deposits in the southern Belize Lagoon. These reefs
grew during the last transgression, and overbank levees, during the last
regression. Carbonate bioherms founded on siliciclastic sediments are present
on the numerous banks near the present shelf edge in the Gulf of Mexico. They
appear to have been formed in the late Pleistocene-Holocene and are on the
topographic highs often associated with salt and/or shale intrusions. In
addition to these biohermal deposits, a blanket of carbonate sediments
(foraminiferal ooze) presently exists over the silica Pleistocene sediments in
the central Gulf of Mexico. This deposit, ranging in thickness from 20 to 50
cm, is the result of a decrease in noncarbonate detrital sedimentation owing to
the cessation of direct detrital transport across the shelf, pelagic deposition
being produced by default.

Adjacent to the Florida Platform, an occasional thick carbonate deposit is
found. These deposits appear to be the direct result of mass transport from the
carbonate shelf to the east. Doyle (1983) described one such feature, which was
the direct result of a large failure on the slope and the resulting deposit at
the base of the scarp. Such activity apparently has occurred frequently
(Freeman-Lynde, 1983). At the extreme southwestern corner of the Florida
platform, a series of canyons appear to have been conduits for the offshelf
transport of shelf carbonates. It is the role of these features and the process
of shelf transport of carbonate material that are the subjects of this report.

2 SOUTH FLORIDA SHELF AND SLOPE

Over 2000 km of high-resolution (800-J) sparker data were collected over
the southwestern corner of the Florida Platform. Surface samples of the
sediment deposits on the shelf were collected by an underway sampler. Specimens
of the reefs and outcrops were obtained by dredging. The sediment section at
the base of the scarp was characterized by analyses of 5-m cores. These data,
with additional sediment data from the literature (Milligan, 1962; Grady, 1971)
form the basis for the characterization of shelf geology and the processes
active on the shelf. Holmes (1984) identified 26 geologic units that
collectively comprise the geology of this part of the Florida Platform (Fig.
1).

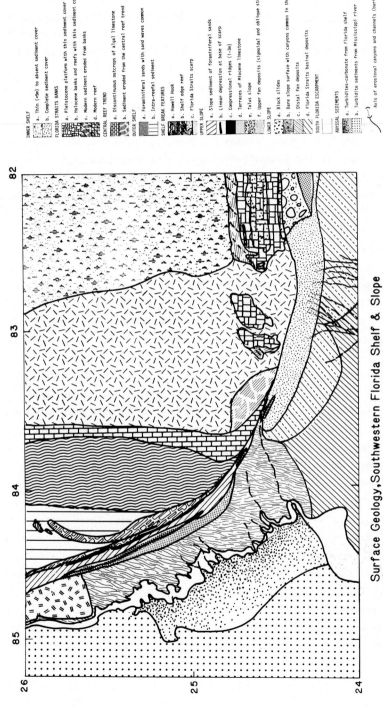

Surface Geology, Southwestern Florida Shelf & Slope

Fig. 1. Geologic map of the southwestern Florida shelf.

INNER SHELF
- a. Thin (<5m) to absent sediment cover
- b. Complete sediment cover

FLORIDA STRAITS BANKS
- a. Pleistocene platform with thin sediment cover
- b. Holocene banks and reefs with thin sediment cover
- c. Modern sediment eroded from banks
- d. Modern reef

CENTRAL REEF TREND
- a. Discontinuous outcrops of algal limestone
- b. Sediment eroded from the central reef trend

OUTER SHELF
- a. Foraminiferal sands with sand waves common
- b. Intra-reefal sediment

SHELF BREAK FEATURES
- a. Howell Hook
- b. Shelf edge reef
- c. Florida Straits scarp

UPPER SLOPE
- a. Slope sediment of foraminiferal sands
- b. Linear depression at base of scarp
- c. Compressional ridges (1-3m)
- d. Terraces of Miocene limestone
- e. Talus slope
- f. Upper fan deposits (sigmoidal and oblique strata)

LOWER SLOPE
- a. Block slides
- b. Bare slope surface with canyons common in the south
- c. Distal fan deposits
- d. Florida Straits basinal deposits

SOUTH FLORIDA ESCARPMENT

ABYSSAL SEDIMENTS
- a. Turbidites-carbonate from Florida shelf
- b. Turbidite sediments from Mississippi river

⌐ Axis of erosional canyons and channels (buried)

2.1 Outershelf

Separating the inner shelf from the outer shelf is a discontinuous north-south-trending, 10-km-wide zone of hardgrounds. This feature is a series of patch reefs that have been partially covered by recent shelf deposition. At present, this trend forms a geomorphic step in the shelf (70 m to 90 m) and is known as Pulley Ridge.

The outer edge of the shelf is marked by a double reef complex. The shallowest reef, cresting from 130 m to 150 m subsea, parallels the shelf break only on the south. The central part of this reef veers landward, forming the feature called Howell Hook (Jordan and Stewart, 1959). Jordan and Steward (1959) and Ballard and Uchupi (1970) speculated that this feature is a spit formed by northward-transported sediment. However, the lack of clinoform bed forms, common in spit development, and an internal structure similar to that of actively growing bioherms (Enos, 1977; Holmes, 1981) suggest that this feature is a bioherm. If this interpretation is correct, the area encompassed is approximately 3300 km^2, and is an ancient analog of a lagoon or a sound similar to Card Sound, on the east coast of Florida.

The lower reef crests at 210 m subsea in the south and 235 m subsea in the north, with a west-facing scarp approximately 40 m high. The crest of the reef is buried by 10 to 15 m of sediment, upon which Howell Hook bioherm rests. The stratigraphic relationship of these two systems has been discussed by Holmes (1984).

Between these reefs and the midshelf trend, the surface of the platform is covered by coarse carbonate sediment. Compositionally, this sediment is biogenic carbonates dominated by foraminifera (50%), with associated mollusca, pterapods, and algal fragments. The pelagic nature of this sediment is attested to by the dominant forms, Globigerinoides trilobus and Globorotalia crassiformis, which are common in Caribbean waters (Jones, 1966). The paucity of fine sediments within these deposits is a result of the strong currents that sweep this section of the shelf. Further evidence of the strong currents is the presence of sand waves, which are common in this area. These features occur in sets of three to five waves with wave heights of 5 m and wave lengths of 0.2 km to 2 km. The asymmetric form of these features, with the steep side to the south, indicates that the currents have a characteristic north-to-south component (Fig. 2). Clinoform beds within the sediment on the extreme southwestern corner of the platform indicate that these features are not recent additions to the shelf but that they have been active throughout the late Holocene.

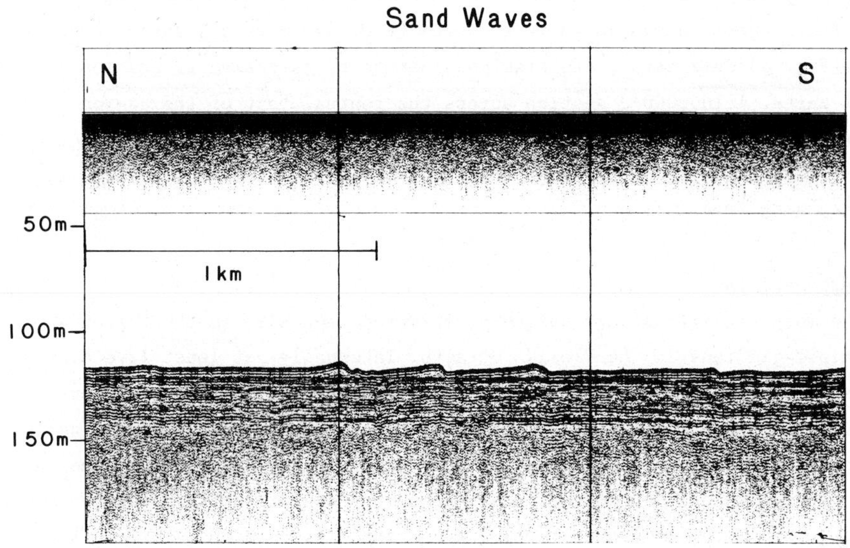

Sand Waves

Fig. 2. Sand wave field on the outer shelf. The waves average 5 m in height and range from 0.2 km to 2 km in length.

2.2 Slope

The upper slope between the shelf break and the terrace at 400 m subsea has a 1° slope inclination. This terrace is approximately 5 km wide and crops out from the extreme southwestern corner of the platform north to approximately 25°45'N, a distance of 120 km. Rocks dredged from the surface are highly phosphatic limestones of Miocene age, encrusted with manganese. The surface of this terrace can be traced to the east under the sediments and reefs that rim the northern Florida Straits and crop out south of the Marquesas Keys as the feature called the Pourtales Terrace. The terrace becomes thinly buried to the north, so that its presence is denoted only by an inflection point on the slope. Below this point the slope is covered with sediment of an undetermined thickness and shows evidence of mass movement. To the south, beyond that point where the terrace is covered (25°45'), the slope is barren of sediment. However, above the terrace on the upper slope, the sediment section also exhibits a form of mass movement. This accordianlike morphology on the surface immediately above the terrace and the tension depression immediately below the shelf-edge reef scarp appear to be a textbook example of mass movement by slumping (Lewis, 1971).

The lower slope in this part of the study area (25°45') is barren of sediment, and the slope is dissected by canyons, which become more numerous to the south. The northernmost features are linear and head below the 1250-m isobath. One canyon system in the south, however, heads above the 100-m isobath and is the largest canyon known in the area (Fig. 3). Sections across this canyon (Fig. 4) demonstrate its erosional nature by extensive truncation of stratal units. A 160,000-J section across the central part of the canyon indicates that this feature has been active for a considerable length of time and has a central fill of sediment that apparently has been derived from the slope and head (Fig. 5).

2.3 Carbonate fans

The most significant and unique periplatform deposits in the region are the hemipelagic fans in the Florida Straits. Internally, at least five cycles of deposition are recognized within the deposits (Fig. 6). The upper four sequences have been correlated to sedimentological events that occurred during the last transgression of the sea, and thus are late Pleistocene-Holocene in age. To the west, the youngest of these sequences are presently being eroded, attesting to the recent canyon-forming processes active in the Florida Canyon (Fig. 7).

2.4 Talus

Immediately above the Pourtales Terrace there is a deposit that on the basis of form and lack of internal structure, is interpreted as a talus slope. An east-west section demonstrates that the talus material is overlapped by the hemipelagic material of the fans. Although detail is lacking, it appears that the upper sequences are present. It is therefore concluded that the talus was deposited when the shelf-edge reefs were active, during the late Pleistocene and prior to the late Holocene.

2.5 Turbidites

At the base of the scarp, distinct turbidite deposits are present (Fig. 8). Cores of this material include some clastics but are dominated by planktonic foraminifera. Compared by composition alone, this deposit would be considered similar to that found in the central Gulf. However, the form and thickness of the deposit demonstrate otherwise. To the west, the thick foraminiferal ooze thins to a thin veneer that covers sediment deposited as turbidites with a Mississippi mineral assemblage. Thus the sedimentary section at the base of the Florida scarp consists of a thick carbonate section over a siliciclastic section derived from Mississippi River turbidites. Alternating

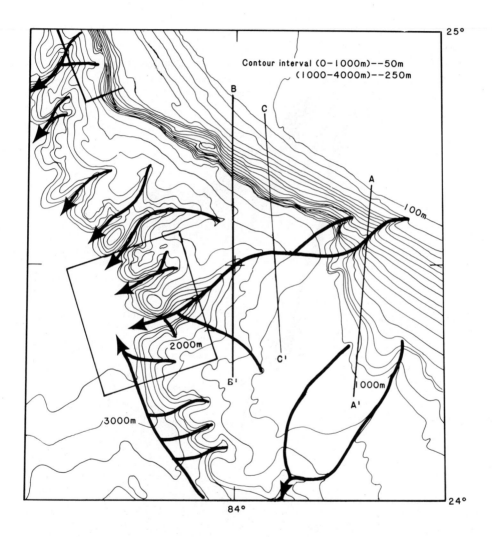

Fig. 3. Bathymetric map of the slope at the southwestern corner of the Florida Platform. The aquare left of center is the area covered by GLORIA in Figure 1. The heavy lines are the thalwags of the canyons cut into the slope.

278

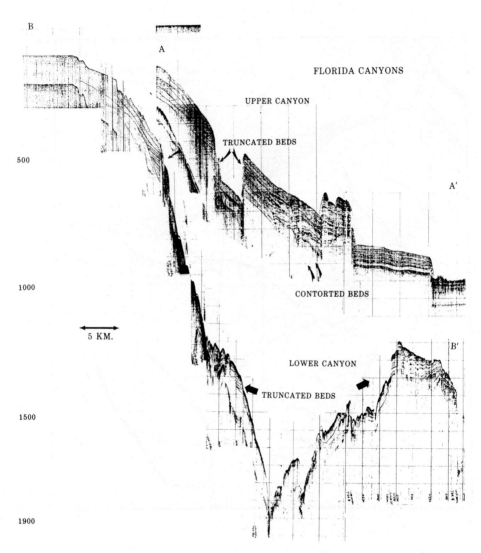

Fig. 4. 800-J seismic profiles across the upper (A-A´) and lower (B-B´) parts of the Florida Canyon. The trancation of the strata is clearly shown in the walls of the canyon.

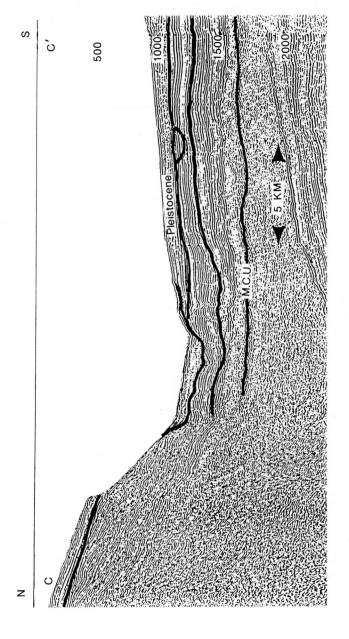

Fig. 5. 160,000-J sparker profile (C-C') across the central part of the Florida Canyon. This profile documents that the canyon has been active since from well into the Pleistocene. The correlations are interpreted from Deep Sea Drilling Program drill sites and seismic tie to this profile.

280

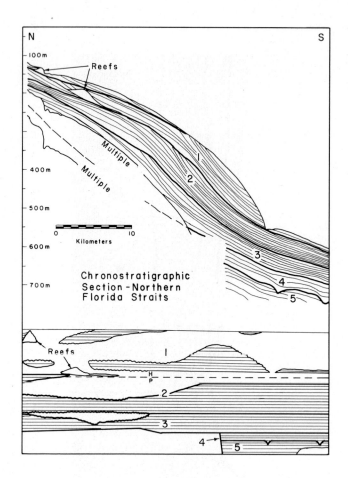

Fig. 6. Chronostratigraphic section of the carbonate fans that extend into the
Florida Straits. Sequences 1-4 have been related to events in the latest
Pleistocene-Holocene transgression. The Holocene-Pleistocene boundary between
sequences 1 and 2 is based on C-14 dates of a reef which has been correlated to
the buried reef resting on the upper surface of sequence 2.

parallel and chaotic reflectors in seismic profiles near the base of the
platform suggest alternating influence of the Mississippi River-derived
material with the carbonate turbidites from the Florida Canyon systems (Fig.
9).

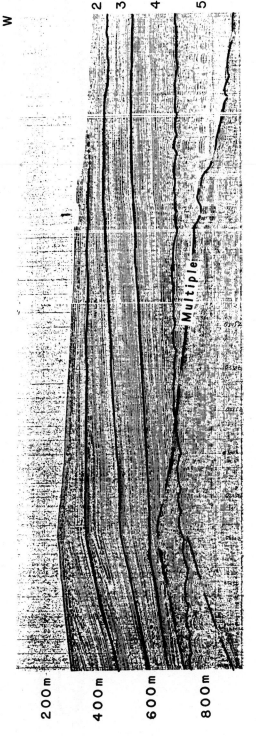

Fig. 7. 800-J sparker E-W profile across the upper part of the fan. Outcropping of the sequences is the result of erosion near the head of the Florida Canyon.

282

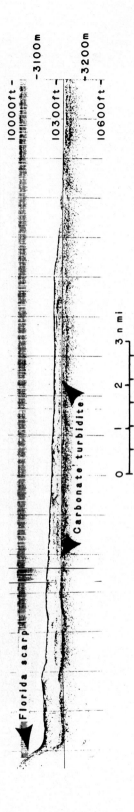

Fig. 8. Carbonate turbidity deposit near the mouth of the Florida Canyon.

283

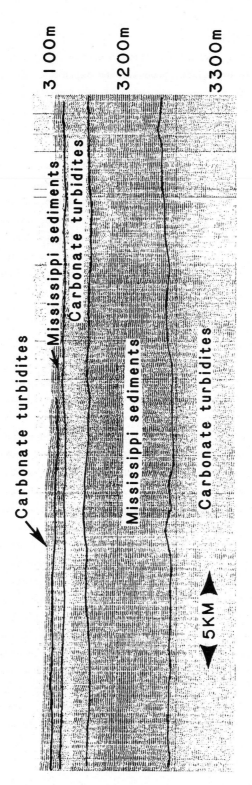

Fig. 9. 800–J sparker profile near the mouths of the canyons.

3 ORIGIN OF CARBONATES

As stated above, the thick carbonate section at the base of the slope is composed of pelagic organisms; however, the depth and form of the deposit indicate a turbidite process as a cause of its deposition, i.e., it is pelagic material derived from a shelf region. Austin (1971) measured the influence of the Loop Current on biologic production in the eastern Gulf. His study defined the areas of high productivity along the Florida shelf edge as resulting from upwelling caused by the impingement of the water mass on the shelf edge. This productivity would increase the production and sedimentation of pelagic forms, and these forms would be characteristic of the Caribbean waters entrained in the Loop Current. Such activity would account for the dominance of Globigerinoide trilobus and Globorotalia crassiformis forms common in Caribbean waters (Jones, 1966). Recent physical oceanographic studies on the shelf have demonstrated that similar upwelling conditions may occur across the outer shelf (Chin, 1983).

The movement of the water mass produces an environment that affects sedimentation in two ways germain to the formation of turbidite deposits. One is due to the circulation of the Loop Current; the Caribbean water would first impinge on the shelf far north of the study area. Because of this, the greatest effect would be near this area; that is, the greatest production and the greatest sedimentation would occur on the central shelf. As the water is transported southward, production and sedimentation would diminish according to the depletion of nutrients. The sum result would be a higher sedimentation rate to the north and would explain the sediment pattern of a thicker cover overriding the terrace to the north. This overriding has caused the sediment to fail and, at one locality, produce a massive failure (Doyle, 1983) of the type described by Cook and Taylor (1977) for the downslope transport of sediment in the Cambrian of western Nevada. Farther to the south, where the slopes are steeper, the high rate of sedimentation has created an unstable sediment deposit, producing downslope creep. In addition, on the shelf the nutrient-depleted water acts to transport sediments, creating sand wave fields.

These conditions are a result of the present ocean circulation and the shelf configuration. In the past, when sea level was lower, oceanic conditions were different. These differences are detailed in the following model relating the sedimentary deposits to one cycle of sea-level transgression, using as an example the late Pleistocene-Holocene transgression. During a slowdown in the last transgression of the sea (18,000 years B.P.), the shelf-edge reef developed. Because the nature of the reef is unknown, the relative position of sea level can only be assumed, but some authors place it at 160-180 m below

present at that time (Shepard and Curray, 1964). It is apparent, then, that a
tectonic readjustment of the shelf occurred, possibly as recently as the
Holocene, which would have caused the shelf-edge reef to be lower relative to
present sea level than during its growth. Evidence of this tectonism is the
drowning of the oolite deposits in the Marquesas region, which are present 3-4
m subsea, but exposed in the islands north of Key West. Assuming synchronity of
these deposits, a tilt of the peninsula is required that is identical to the
present slope of the Miocene surface (Holmes, 1981). At the time of this reef
growth, the major sedimentation in the basin was wholly terrigenous from the
Mississippi River system because of the volume of water flowing into the Gulf
from glacial melt. Such conditions reduced the strength of the Loop Current
negating its effect in producing pelagic sediment by upwelling.

Approximately 10,000 years B.P. the sea surface was between 80 and 110 m
below present levels (Milliman and Emery, 1968). This time is considered to be
the break between the Pleistocene and the Holocene (J. Tracey, pers. commun.,
1982) and appears to coincide with an interstadial within the last transgression. During this period, Howell Hook was formed and shelf sedimentation
was initiated. The sedimentation partially buried the lower reef, and much of
the material was transported along the shelf south to the canyons. This process
initiated conditions responsible for turbidite sedimentation, which was
becoming more dominant owing to the decline in Mississippi River influence.

In the middle Holocene (~5000 years B.P.), with sea level within a few
meters of the present level, major sedimentation sites were present on the
shelf and slope. The water level had breached the gaps between the southern
bank to allow sediment to be transported and deposited in the Florida Straits.
Even though the Florida current is very strong, the main axis apparently was
far to the south, so that the loci of deposition occurred where very little
current existed.

The final step, with sea level attaining its present level, resulted in
cessation of sedimentation on the fans as a result of east-west tidal exchange
through Florida Bay, possibly with an added water head caused by precipitation
on the Florida Peninsula. This exchange has deflected the shelf current to the
west around the southern banks resulting in delivery of the shelf sediment to
the canyon heads and thence to the basin flow as turbidite deposits. These
conditions produced the major canyon whose head reaches up to the 100-m
isobath, at the junction of the Yucatan Channel and Florida Straits.

This canyon was initially formed by slumping in a manner conceptualized by
Farre et al. (1983), with further development by the downward transport of
shelf material. As the head of the canyon worked landward, it eroded to the

Tertiary rocks that make up the terrace. These densely consolidated strata deflected the headward development to the east, so that at present the upper section of the canyon follows the trend of the buried Tertiary shelf edge.

Because this part of the canyon cuts late Pleistocene and Holocene sediments that were deposited in the carbonate fans, the canyon is seen as modern, with the sediment being fed into the system from erosion and deposition of material transported along the shelf.

In conclusion, the alternating carbonate and siliciclastic sediments found at the base of the Florida Platform are a product of the oceanic condition differential between maximum and minimum sea levels. These deposits, although composed of planktonic organisms, were deposited as turbidites. The only information as to their origin is the geometric form of the deposit, with an occasional clastic sediment eroded from the canyon walls. In the geologic record, these deposits would be intercalated limestones and shales. It would be difficult to differentiate these deposits from a pelagic-detrital depositional environment. Differentation would require detailed petrographic examinations of the carbonate to ascertain the amount of nonpelagic material present.

4 REFERENCES

Austin, H., 1971. The characteristics and relationship between calculated geostrophic current components and selected indicator organisms in the Gulf of Mexico Loop Current system; Ph.D. Dissertation, Dept. of Oceanography, Florida State University, 369 pp.

Ballard, R.D. and Uchupi, E., 1970. Morphology and Quaternary history of the Continental shelf of the Gulf Coast of the United States. Bulletin of Marine Science, 20, 3, 547-559.

Chin, H., 1983. Southwest Florida continental shelf: A Loop Current mechanism for productivity enhancement and questions for further study. Proceedings, Winter Ternary Gulf of Mexico Studies Meetings, Minerals Management Studies, P. 33-38.

Choi, D.R. and Holmes, C.W., 1983. Foundations of Quaternary reefs in south central Belize lagoon: American Association Petroleum Geologists Bulletin, 66, 263-2671.

Cook, H.E., Taylor, M.E., 1977. Comparison of continental slope and shelf environments in the Upper Cambrian and Lower Ordovician of Nevada in Cook, H.E., and Enos, P., Deep Water Carbonate Environments, S.E.P.M. Special Publication No. 25, p. 51-81.

Doyle, L.J., 1983. Shallow structure and stratigraphy of the carbonate west Florida continental slope and their implications to sedimentation and geohazards: U.S. Geol. Survey Open-File Report 83-425, 19p.

Enos, P., 1977. Holocene sediment accumulations of the South Florida shelf margin. In: P. Enos and R. Perkins, Quaternary sedimentation in South Florida: Geological Society of America Memoir 147, pp. 1-130.

Farre, J.A., McGregor, B.A., Ryan, W.B. and Robb, J.M., 1983. Breaching the shelf break: Passage from youthful to mature phase in canyon evolution. In: D.J. Stanley and G.T. Moore (Editors), The Shelf Break: Critical Interface on Continental Margins, S.E.P.M. Special Publication 33, pp. 25-40.

Freeman-Lynde, R.P., 1983. Cretaceous and Tertiary samples dredged from the Florida escarpment, eastern Gulf of mexico. Transactions, Gulf Coast Association Geological Societies, Jackson, 1983, 33.

Grady, J.R., 1971. Tistribution of sedimentary properties and shrimp catch on two shrimping grounds on the continental shelf of the Fulf of Mexico. Proceedings, Gulf, Caribbean Fish Institute, 23rd Annual Session, p. 139-148.

Holmes, C.W., 1973. Distribution of selected elements in the surficial sediments of the northern Gulf of Mexico continental shelf and slope. U.S. Geological Survey Prof. Paper 813, 9 p.

Holmes, C.W., 1981. Late Neogene and Quaternary geology of the South Florida shelf and slope. U.S. Geological Survey Open-File Report 81-79.

Holmes, C.W., 1984. Accretion on the South Florida Platform, late Quaternary development, in press.

Jones, J., 1966. Distribution and variation of living pelagic froaminifera in the Caribbean Sea Proceedings, 3rd Carib. Geol. Conf., p. 178-183.

Jordan, G.F. and Steward, H.B., Jr., 1959. Continental slope off southwest Florida. American Association Petroleum Geologists Bulletin, 43, 974-991.

Lewis, K.B., 1971. Slumping on a continental slope incline 1°-4°. Sedimentology, 16, 97-110.

Milligan, D.B., 1962. Marine Geology of the Florida Straits, M.S. Thesis, Oceanog. Institute, Florida, Florida State University., 130p.

Milliman, J.D. and Emery, K.O., 1968. Sea levels during the past 35,000 years. Science, 4, 167, p. 1121-1123.

Shepard, F.P. and Curray, J.R., 1964. Carbon-14 determinations of sea level changes in stable areas. Progress in Oceanography, 4, 283-291.

Chapter 11

CONTROL OF TERRIGENOUS-CARBONATE FACIES TRANSITIONS BY BAROCLINIC COASTAL
CURRENTS - NICARAGUA

S.P. MURRAY, H.H. ROBERTS and M.H. YOUNG
Coastal Studies Institute, Louisiana State University, Baton Rouge,
Louisiana 70803

ABSTRACT: The cross-shelf movement of fine-grained terrigenous sediment has
received considerable attention in recent geological literature. Of perhaps
greater interest is the spatial relationship with other sediment types and
the processes that control facies segregation. Studies of sediment
distribution on the shallow and broad shelf off the east coast of Nicaragua
have revealed a 20- to 30-km-wide (12- to 19-mi) band of terrigenous
sediment confined near the coast. The band grades abruptly into an area of
carbonate deposition that is composed principally of the disintegration
products of calcareous green algae.
    Detailed studies of watershed runoff, structure of nearshore currents, and
density gradients indicate the existence of a well-organized band of
currents that run essentially parallel to the coast. An analytical model is
formulated that successfully predicts the strength and location of this
coastal boundary current system as a function of wind speed and direction,
water depth, and local density gradients. The analytical prediction of the
cross-shore current structure agrees well with the field observations in
delineating the presence of convergences and divergences that will act to
restrict the seaward migration of fine-grained terrigenous particulates and
thus preserve the identity of the two contiguous facies. Variations in the
model inputs (e.g., rainfall runoff, wind regime) allow estimates of how
such systems operated in different geological regimes, and consequently how
such facies relations evolved in climatic regimes different from that of the
present time. An understanding of the physical processes that control the
siliciclastic-carbonate interface on the eastern Nicaragua shelf emphasizes
the delicate balance that exists between terrigenous and carbonate dominance
over the shelf.

1 INTRODUCTION

    Recent work (Royer, 1981; Blanton and Atkinson, 1983) has emphasized the

importance of freshwater inflow to the coast in forming high-speed nearshore

current streams that run along many coasts. Both Blanton (1981) and Murray

et al. (1982) note that these types of nearshore currents could become

important unidirectional conduits for longshore transports of suspended

material of terrigenous origin and pollutants introduced into coastal waters

by river flow.

    The cross-sectional distribution of mean velocity within the current

itself remains a problem of significant interest because the structure of

the mean flow is heavily masked in most areas studied thus far by strong

tidal currents. Unusually steady trade winds and negligible tidal currents

along the east coast of Nicaragua allow a detailed look at this velocity
structure and the role it plays on the dispersal of sediment on the
underlying and adjacent shelf.

The purpose of this paper is to determine the principal driving forces in
the nearshore current off Nicaragua and to demonstrate the applicability of
a simple analytical model that describes the primary characteristics of the
cross-sectional velocity distribution. The model is then used to explain the
surprising dominance of carbonate deposition and reef growth on a shelf that
is receiving an abnormally large volume of terrigenous sediment.

2 BACKGROUND

Because of the large amount of rainfall (>6 m year^{-1}) on coastal
watersheds on the east coast of Nicaragua, a freshwater discharge from 11
major rivers estimated at 1.45×10^{11} m^3 yr^{-1} reaches the coast along the
450-km stretch from the major turn in the coast at Cape Gracias a Dios in
the north to the Costa Rican border in the south, where the broad shelf
terminates (Figure 1). This turbid discharge, carrying an estimated total
sediment supply of $24-32 \times 10^6$ metric tonnes/year, acts as a line source to
produce a continuous band of brackish water that moves southerly along the
coast. Stratified in salinity but isothermal, it typically extends 20-30 km
offshore. As part of a larger study of the meteorology, physical
oceanography, and sedimentation along the east coast of Nicaragua, a number
of velocity and temperature-salinity observations were made across the shelf
(Figure 1). The physical oceanography and a preliminary analysis of the
dynamics were given in Murray et al. (1982). Roberts and Murray (1983)
described the sedimentological facies across the entire shelf.

This paper will focus on a theoretical model that predicts the
cross-sectional distribution of velocity through the band of nearshore
current out to the edge of the shelf regime proper and explains how the
predicted velocity field controls the sharp transitions observed in the
sedimentological facies.

The alongshore and cross-shore velocity components from our current meter
observations along a line near El Bluff (Figure 1), presented in Figures 2
and 3, respectively, are representative of the entire coast. The alongshore
speed (Figure 2) shows a well-defined maximum at about 11 km offshore,
diminishing to zero at about 18 km, beyond which a slow northerly
countercurrent appears. This structure is clearly intimately related to the
density (salinity) field, also plotted in Figure 2.

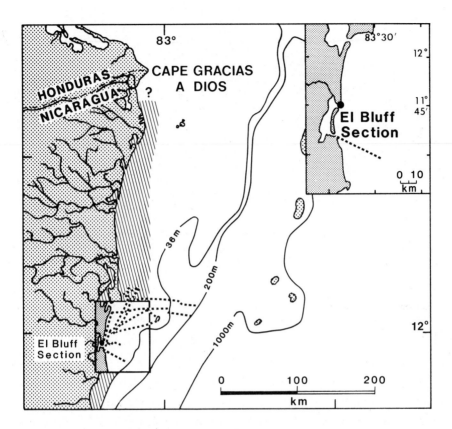

Fig. 1. Index map of study area along the Atlantic coast of Nicaragua.

The cross-shore flow component (Figure 3) shows a strong surface convergence and a bottom divergence near the longshore current maximum at kilometer 12. The upper layer water appears to sink gradually to the bottom as it moves offshore, providing a conduit seaward for the near-coastal water.

Tides in the area are small, with amplitudes of 5-8 cm and tidal currents (4-8 cm s^{-1}) an order of magnitude smaller than the speeds in the core of the nearshore current. Most of the year, winds in this trade wind zone are remarkably steady at 7-10 m s^{-1} from the ENE, the steadiness factor (ratio of vector mean wind to mean wind speed) being above 90% [Murray et al. (1982)].

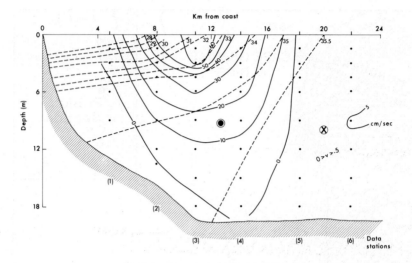

Fig. 2. Distribution of the alongshore component of the current velocity (centimeters per second) and salinity (parts per thousand) along the El Bluff section.

3 FORCES AT WORK

The usual equations (e.g., Pond and Pickard, 1978) describing the balance between the accelerations of the water column (on the left side of Eq. 1) and the forces acting on it (the right side of Eq. 1) are:

$$\frac{du}{dt} = fv - \frac{g}{\rho} \int_o^z \frac{\partial \rho}{\partial x} \, dz - g \frac{\partial \eta}{\partial x} - \frac{1}{\rho} \frac{\partial \tau_x}{\partial z} \tag{1a}$$

$$\frac{dv}{dt} = -fu - \frac{g}{\rho} \int_o^z \frac{\partial \rho}{\partial y} \, dz - g \frac{\partial \eta}{\partial y} - \frac{1}{\rho} \frac{\partial \tau_y}{\partial z} \tag{1b}$$

where the hydrostatic approximation to the vertical momentum equation is employed and only vertical transfer of momentum by eddy friction is considered. These forces, from left to right, are: the Coriolis forces, the forces arising from the density gradient, the water surface slope forces, and the forces of internal friction are all represented by conventional notations. For example, g is the acceleration of gravity and u and v are current velocity components in the x and y directions, respectively. We take z positive down, η the sea surface coordinate positive up from the mean level, h the coordinate of the bottom, x (u) positive in the direction

293

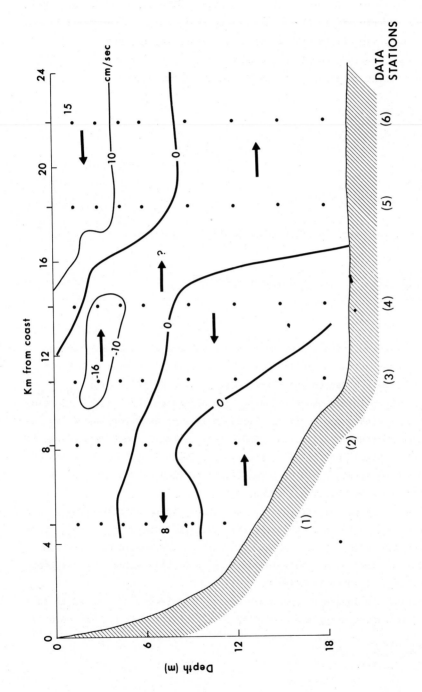

Fig. 3. Distribution of the cross-shore component of the current velocity (centimeters per second) along the El Bluff section.

toward the shore, and y(v) positive in the direction counterclockwise of positive u (to the left of an observer looking onshore). For the data under study here, Murray et al. (1982) have shown that, owing to the small tide and the steady wind field, the local acceleration and the field accelerations are in general negligible, i.e., $du/dt \simeq dv/dt \simeq 0$.

The vertically integrated continuity equation reduces to

$$\frac{\partial q_x}{\partial x} + \frac{\partial q_y}{\partial y} = 0 \qquad (2)$$

where

$$q_x = \int_{-\eta}^{h} u \ dz, \quad q_y = \int_{-\eta}^{h} v \ dz.$$

Integrating (1a) over the water depth yields an expression

$$\frac{\partial \eta}{\partial x} = \frac{1}{gh} \int_{-\eta}^{h} (fv - \frac{g}{\rho} \int_{-\eta}^{z} \frac{\partial \rho}{\partial x} \ dz) \ dz + (\frac{\tau_{xs} - \tau_{xh}}{\rho g h}) \qquad (3)$$

which allows computation of the cross-shore surface slope $\partial \eta / \partial x$ from the observed values of the Coriolis term, the baroclinic pressure gradient term, and the surface and bottom shear stresses, $\tau_{xs} = \kappa \ \rho_a \ |W| \ W_x$ and $\tau_{xh} = C_d$ $\rho |V_h| \ u_h$, respectively, where κ and C_d both are taken $\simeq 2.5 \times 10^{-3}$ (Bowden et al., 1959; Hickey and Hamilton, 1980), W is wind velocity, and V_h is the near-bottom current velocity. The wind stress vector for our data on the El Bluff line is 0.23 dynes/cm^2 directed along azimuth 203° true. Solution of (3) then allows substitution of $\partial \eta / \partial x$ back into (2a), which can then be solved for the eddy frictional force, τ_x, τ_y.

The computed internal stresses τ_{xz} are then used to estimate eddy viscosities N_{xz} in the section using the defining relation $\tau_{xz} = -N_{xz} \ \partial u / \partial z$. Of 23 estimates of N, 80% are <8 cm^2 s^{-1} with no systematic distribution across the section and a mean value of 6 cm^2 s^{-1}. This value is consistent with previous estimates (Pettigrew, 1981).

Neglecting the longshore density gradient term $[0(10^{-6})$ compared to fu $= 0(10^{-4})]$, and integrating and averaging over the vertical, (1b) becomes

$$\frac{\partial \eta}{\partial y} = - \frac{f \bar{u}}{g} + \frac{(\tau_{ys} - \tau_{yh})}{g \rho h} \qquad (4)$$

which allows an estimate of the longshore surface slope ($\bar{u}$ is a vertical average). The four values computed from the El Bluff section have an average value $\partial \eta / \partial y \approx 6 \times 10^{-8}$ which should be representative of a general slope of the nearshore band down to the north against the wind.

4   THE CROSS-STREAM VELOCITY DISTRIBUTION

4.1 Formulation

For the predictive model, certain simplifying approximations are necessary. First, the density gradient $\partial p / \partial x$ is not depth dependent, an assumption that is violated only in the lower regions of the cross section, where velocities are minimal and will have the least effect on the primary flow pattern. Second, the eddy viscosity N is likewise independent of depth. Considering the wide range of eddy viscosities in nature, this is not an unreasonable assumption for our data on N, discussed earlier.

Combined with the analysis in the preceding section, the controlling momentum equations (1a, b) then reduce to

$$0 = fv - \frac{z + \eta}{\rho} \frac{\partial p}{\partial x} - g \frac{\partial \eta}{\partial x} + N \frac{\partial^2 u}{\partial z^2} \qquad (5a)$$

$$0 = -fu - g \frac{\partial \eta}{\partial y} + N \frac{\partial^2 v}{\partial z^2} \qquad (5b)$$

If the discharge q_y is uniform along the coast, the continuity equation (2) becomes q_x = constant, usually either $q_x = 0$ or $q_x = -q_F$, where q_F is an equivalent line source discharge of fresh water to the coast (Heaps, 1972).

Defining the complex velocity W = u + iv, (5a, b) can be put in the complex form

$$\frac{\partial^2 W}{\partial z^2} = \alpha^2 W - \alpha^2 \frac{ig}{f} \{ (z + \eta) \frac{1}{\rho} \frac{\partial p}{\partial x} + S \} \qquad (6)$$

where S is the complex slope = $\partial \eta / \partial x + i (\partial \eta / \partial y)$ and $\alpha^2 = if/N$.

The solution to (6) is

$$W = A \exp \alpha(z + \eta) + B \exp - \alpha(z + \eta)$$

$$+ \frac{ig}{f} \{ \frac{z + \eta}{\rho} \frac{\partial p}{\partial x} + S \} \qquad (7)$$

where A and B are complex constants to be determined by the boundary conditions. The cross-shore volume flux is given by the real part of

$$\int_{-\eta}^{h} W \, dz = \frac{A}{\alpha} (e^{\alpha h} - 1) - \frac{B}{\alpha} (e^{-\alpha h} - 1)$$

$$+ \frac{igh}{2\rho f} (\frac{\partial \rho}{\partial x} h + 2\rho S) \tag{8}$$

where terms in η are neglected as usual, as they affect the transport by only $O(0.1\%)$. The model differs from that of Heaps (1972) by the inclusion of wind stress, quadratic bottom friction, and a longshore surface slope.

The complex constants A and B and the cross-shore slope $\partial \eta/\partial x$ are considered unknown in (7). From the preceding dynamical analysis $\partial \eta/\partial y$ is taken $\simeq 5.7 \times 10^{-8}$. Although negligible here, the solution can also include an along shore baroclinic pressure gradient.

The surface wind stress boundary condition $\partial W/\partial z = - \tau_s/N$, where τ_s is complex, determines the first complex constant

$$A = B - \frac{ig}{\alpha f \rho} \frac{\partial \rho}{\partial x} - \frac{\tau_s}{\alpha N} \tag{9}$$

The bottom stress boundary condition

$$N(\frac{\partial W}{\partial z})_h + \kappa \rho |W_h| W_h = 0 \tag{10}$$

and the continuity equation close the system by allowing computation of the constant B and the cross-shore slope $\partial \eta/\partial x$. Further details are given in Murray and Young (1985).

The vertical current profile u and v can now be predicted from (7) using observed or estimated values of the wind stress, eddy viscosity, water depth, cross-shelf density gradient, and longshore surface slope.

4.2 Results of Computations

The distribution of the alongshore velocity component calculated from (7) and (8) under the wind stress, density gradients, and eddy viscosities associated with the field data presented in Figure 2 is given in Figure 4. Although minor details deviate, the agreement between theory and observation is extremely encouraging. For example, the magnitude and location of the current maxima at 12 km offshore and the offshore length scale of 15-20 km are all successfully reproduced. The offshore countercurrent of 5-10 cm s^{-1} in both observations and theory is apparently the result of the slight longshore surface slope opposing the wind.

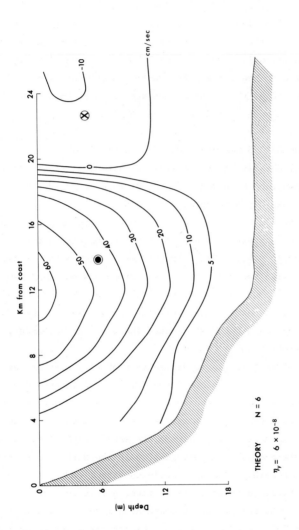

Fig. 4. Distribution of the alongshore component of the current velocity as predicted by the analytical model.

Comparison of the cross-shore velocity field from theory (7) (Figure 5) with its observational counterpart (Figure 3) points up several important similarities. Seaward of kilometer 16 a fairly strong (5-15 cm s^{-1}) onshelf-directed upper layer is underlain by a thicker, weaker (5 cm s^{-1}) offshelf-directed lower layer. Landward of approximately kilometer 16 the situation reverses, with an offshelf-directed upper layer underlain by an onshelf-directed lower layer. A convergence intersects the surface at about kilometer 12 in both theory and observation. This convergence will force sediment-laden near-surface water from the onshore side of the convergence to plunge downward to greater depths. Particulates that have settled out to a depth 12 meters below the surface then appear to be trapped within the nearshore zone by the onshore-directed bottom layer.

5 CONTROL ON TERRIGENOUS-CARBONATE FACIES DISTRIBUTION

The structural details of the coastal current discussed in detail above are extremely important with regard to understanding the distribution of sediments on the eastern Nicaragua shelf. Additionally, estimates of southerly sediment flux by the coastal current, based on actual observations of suspended load, suggest a transport of approximately 15.8×10^{6} metric tonnes/year (Murray et al., 1982). The ultimate fate of this sediment is largely unknown, with the exception of a terrigenous sediment ramp that has been constructed on the nearshore shelf (Roberts and Murray, 1983). It is probable that a large part of the sediment load carried by the coastal current is transported off the shelf near the Costa Rican border, where the coast changes orientation.

A rapid shift from terrigenous sediments near the coast to carbonates in a shelfward direction reflects confinement of the sediment-laden brackish water to the inner shelf as dictated by the coastal current. Figure 6 clearly illustrates this distinct separation of sedimentary facies. The excursion of terrigenous sediments onto the shelf south of Punta Perlas (Figure 6) is interpreted as a perturbation in the north-to-south-running coastal current, possibly an eddy associated with flow separation occurring at this abrupt change in coastal orientation, a feature observed by Crout and Murray (1979). Bottom sediments containing more than 40% terrigenous components are rarely found more than 20 km offshore, and in most cases sediments of this description are found in a coast-parallel belt no more than 10 km wide. Sediments deposited from the water column in this zone tend to be trapped near shore by onshore-directed currents, predicted by the model (Figure 5). The core of the alongshore current jet (>50 cm/s), as

299

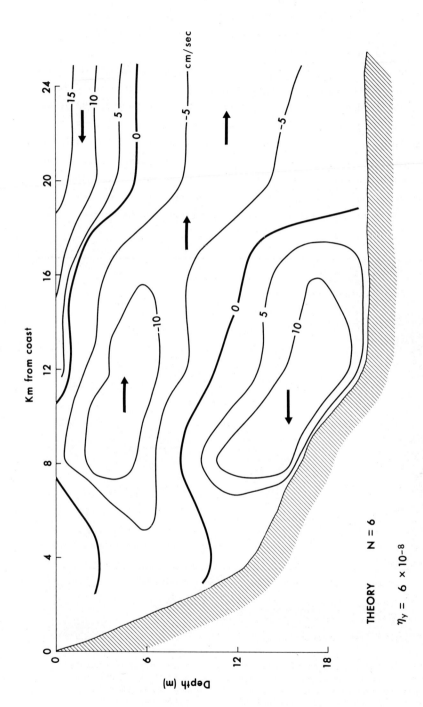

Fig. 5. Distribution of the cross-shore component of the current speed as predicted by the analytical model.

shown in Figure 4, also helps limit the shelfward migration of turbid coastal water by quickly deflecting riverine effluent plumes to an orientation nearly parallel to the coastline.

Echo-sounder profiles, side-scan sonar data, and bottom samples confirm a rapid change in facies and a systematic change in bottom topography from the coast to the shelf edge (Murray et al., 1982; Roberts and Murray, 1983). As Figure 6 shows, a depositional ramp composed primarily of terrigenous sediments is being deposited in a zone from the shoreline to 5-8 km offshore (Owens and Roberts, 1979). Sediment properties in this nearshore region vary considerably with proximity to major sites of riverine input. Accumulations of poorly sorted, coarse volcanic sand are common adjacent to and downdrift of the major river mouths. Better sorted sands compose the beach-bar

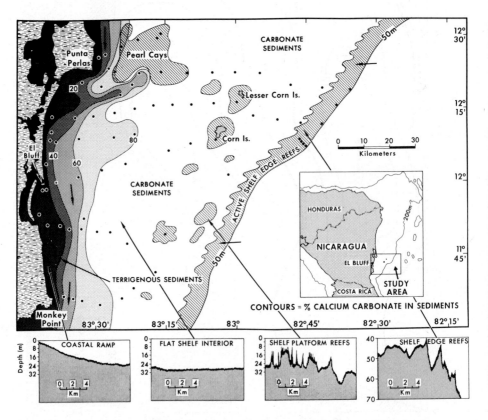

Fig. 6. Contoured percentages of calcium carbonate in bottom sediments. Dots represent sample locations. Hatched areas indicate the location of mid-shelf platforms on the shelf margin belts, where coral reefs are actively growing. Typical echo sounder profiles of the various topographical provinces of the shelf are illustrated.

systems. These nearshore sand bodies grade quickly offshore into silty clays
and clays. At the toe of the ramp (approximately 6-10 km offshore),
terrigenous sediments are progressively diluted in a seaward direction with
poorly sorted carbonate particles.

The relatively flat bottom of the central shelf (Figure 6) is unaffected
by sedimentation processes near the coast. This central shelf is a vast
region of carbonate sediment deposition (water depths - 30-35 meters) where
abundant growth of <u>Halimeda</u> produces a coarse carbonate sediment (Figure 7).
These and other coarse components (mollusc shells, coral fragments, etc.)
are less commonly found in a matrix of aragonitic mud, which also represents
a disintegration product of calcareous green algae.

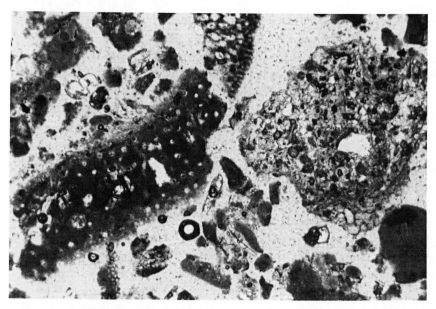

Fig. 7. A thin-section photomicrograph of mid-shelf sediment illustrates the
abundance of (a) <u>Halimeda</u> as well as (b) large foraminifera (<u>Peneroplis</u>?)
fragments. Scale bar = mm.

6 DISCUSSION AND CONCLUSIONS

The dynamical analysis shows that the density gradients arising from the
introduction of fresh riverine water to the nearshore shelf accounts for
upward of 80% of the momentum in the coastal current. That is, the current
is essentially in geostrophic balance. Direct driving of the unusually
steady and nearly unidirectional trade wind plays the critical role of

302

maintaining the intensity of the density gradients through the convergence
of a wind-driven surface Ekman layer toward the coast.

Although the core of this current has been observed to meander like a
river, these lateral excursions rarely extend beyond 20 km from the coast.
Only when relaxation of the constant onshore wind stress occurs or when the
wind blows briefly from a westerly quadrant can significant volumes of
turbid, brackish coastal water migrate onto the shelf. In the yearly wind
cycle there is a brief slackening of the trades in April-May. This period
(including March) also coincides with the period of minimum rainfall, thus
reducing runoff and subsequent impact of terrigenous sedimentation on the
shelf. Although these short-term changes in the coastal current do occur,
sediment distribution and, to a lesser extent, shelf morphology reflect the
long-term dynamic structure of the shelf waters.

In the perspective of geologic time a migration of the trade wind belt
could decrease the wind stress required to maintain the coastal current,
leading to an expansion and thickening of the terrigenous facies in a
shelfward direction. A similar sedimentologic response would occur if
shifting climatic patterns produced a wind field with southerly and westerly
components. Due to rotation of the earth, wind from the south would tend to
deflect coastal water offshore, breaking down the coastal current and
dispersing terrigenous sediment seaward or to the east. Along a modern coast
this type of reversing wind cycle has been documented to occur at the annual
frequency off the South Carolina coast by Blanton and Atkinson (1983), who
correlated the rate of freshwater loss from the coastal current with the
seasonally reversing longshore wind.

An important result suggested by successful comparison of the theoretical
model to field data on currents and sediment distribution is that reasonable
predictions of nearshore currents and patterns of sedimentation can be made
based on estimates of local wind (speed and direction), sediment and fresh
water supply to the coast, and bathymetry. Perhaps this approach can be
exploited as a quantitative technique in reconstruction of certain
paleoenvironments.

More specifically, we conclude:

1. Extreme rainfall and rapid erosion of volcanic rocks under tropical
conditions result in the transport of larger volumes of fresh water and
sediment to the eastern Nicaragua nearshore shelf.

2. Density gradients set up from riverine effluents and maintained by
wind stress provided by the persistent northeast trades are the key elements
in producing a strong (speeds exceeding 70 cm/s) north-to-south-flowing
coastal current.

3. A theoretical model successfully predicts the longshore coastal current and especially the cross-shore velocity components that tend to trap the terrigenous sediment within 10-20 km of the coast.

4. The distribution of terrigenous sediments on the Nicaraguan shelf is controlled by the presence and behavior of the coastal current. Therefore, the zone of terrigenous sediment influence is no wider than the band of southward-moving coastal water. The remainder of the shelf has developed into a vast carbonate province behaving as if the abundant riverine discharge along the coast did not exist. The facies shift from nearly 100% terrigenous sediment to nearly 100% carbonates occurs over a distance generally less than 20 km.

ACKNOWLEDGEMENTS

We gratefully acknowledge support of our studies by the Office of Naval Research, Coastal Sciences Program, Arlington, Virginia 22217.

REFERENCES

Blanton, J.O., 1981. Ocean currents along a nearshore frontal zone on the continental shelf of the southeastern United States. Journal of Physical Oceanography, 11, pp. 1627-1637.

Blanton, J.O. and Atkinson, L.P., 1983. Transport and fate of river discharge on the continental shelf of the southeastern United States. Journal of Geophysical Research, 88 (C8), pp. 4730-4738.

Bowden, K.F., Fairbairn, L.A. and Hughes, P., 1959. The distribution of shearing stresses in a tidal current. Geophysical Journal, 2(4), pp. 288-305.

Crout, R.L. and Murray, S.P., 1979. Shelf and coastal boundary currents, Miskito Bank of Nicaragua. Proceedings, Sixteenth Coastal Engineering Conference (Hamburg). American Society of Civil Engineers, New York, pp. 2715-2729.

Heaps, N.S., 1972. Estimation of density currents in the Liverpool Bay area of the Irish Sea. Geophysical Journal of the Royal Astrological Society, 30, pp. 415-432.

Hickey, B.M. and Hamilton, P., 1980. A spinup model as a diagnostic tool for interpretation of current and density measurements on the continental shelf of the Pacific Northwest. Journal of Physical Oceanography, 10, pp. 12-24.

Murray, S.P., Hsu, S.A., Roberts, H.H., Owens, E.H. and Crout, R.L., 1982. Physical processes and sedimentation on a broad shallow bank. Estuarine, Coastal and Shelf Science, 14, pp. 135-157.

Murray, S.P. and Young, M.H., 1985. (in press) The nearshore current along a high-rainfall trade-wind coast--Nicaragua, Coastal, Estuarine and Shelf Science.

Owens, E.H. and Roberts, H.H., 1979. Variation of wave energy levels and coastal sedimentation, eastern Nicaragua, Proceedings, Sixteenth Coastal Engineering Conference (Hamburg). American Society of Civil Engineers, New York, pp. 1195-1214.

Pettigrew, N.R., 1981. The dynamics and kinematics of the coastal boundary layer off Long Island. Ph.D. thesis, Massachusetts Institute of Technology/Woods Hole Oceanographic Institute, WHOI-81-14, p. 262.

Pond, S. and Picard, G.L., 1978. Introductory Dynamic Oceanography. Pergamon Press, New York, p. 241.

Roberts, H.H. and Murray, S.P., 1983. Control of reef development and the terrigenous-carbonate interface on a shallow shelf, Nicaragua (Central America). Coral Reefs, 2, pp. 71-80.

Royer, T.C., 1981. Baroclinic transport in the Gulf of Alaska, part II, A freshwater driven coastal current. Journal of Marine Research, 9, pp. 251-266.